Student Solutions Manual

for
McKeague's

Basic Mathematics

Sixth Edition

Sheryl W. Sippel
Hillsborough Community College

THOMSON

BROOKS/COLE

Australia • Canada • Mexico • Singapore • Spain • United Kingdom • United States

For more information about our products, contact us at:
Thomson Learning Academic Resource Center
1-800-423-0563

For permission to use material from this text or product, submit a request online at
http://www.thomsonrights.com.
Any additional questions about permissions can be submitted by email to **thomsonrights@thomson.com.**

Thomson Higher Education
10 Davis Drive
Belmont, CA 94002-3098
USA

Asia (including India)
Thomson Learning
5 Shenton Way
#01-01 UIC Building
Singapore 068808

Australia/New Zealand
Thomson Learning Australia
102 Dodds Street
Southbank, Victoria 3006
Australia

Canada
Thomson Nelson
1120 Birchmount Road
Toronto, Ontario M1K 5G4
Canada

UK/Europe/Middle East/Africa
Thomson Learning
High Holborn House
50–51 Bedford Road
London WC1R 4LR
United Kingdom

Latin America
Thomson Learning
Seneca, 53
Colonia Polanco
11560 Mexico
D.F. Mexico

Spain (including Portugal)
Thomson Paraninfo
Calle Magallanes, 25
28015 Madrid, Spain

TABLE OF CONTENTS

Chapter 1 Pretest

1. $7{,}062 = 7{,}000 + 60 + 2$

3. Eighteen thousand, five hundred seven $=$ 18,507

5.
$$\begin{array}{r} \overset{1\ 1}{1{,}029} \\ +\ 4{,}381 \\ \hline 5{,}410 \end{array}$$

7. Round 6,798 to the nearest hundred:
locate the digit to the right of the hundreds:
6,7$\underline{9}$8: 9 ≥ 5 so replace it and all digits to
the right with a 0 and add one to the
hundreds place.   6,800

9.
$$\begin{array}{r} \overset{6\ \ 9\ \ 10}{1{,}7\cancel{0}\cancel{0}\cancel{0}} \\ -1{,}4\,3\,6 \\ \hline 2\,6\,4 \end{array}$$

11.
$$\begin{array}{r} 536 \\ \times\ 40 \\ \hline \end{array}$$
$000\ \leftarrow 0(536)$
$\underline{21{,}440}\ \leftarrow 40(536)$
$21{,}440\ \text{Add}$

13.
$$\begin{array}{r} 174 \text{ R}16 \\ 23\overline{)4{,}018} \\ \underline{23\downarrow} \\ 171 \\ \underline{161\downarrow} \\ 108 \\ \underline{92} \\ 16 \end{array}$$

15. $P = 2l + 2w$
$= 2(36\text{ in}) + 2(27\text{ in})$
$= 72\text{ in} + 54\text{ in}$
$= 126\text{ in}$

$A = lw$
$= (36\text{ in})(27\text{ in})$
$= 972\text{ in}^2$

17. $7 + 3 \cdot 2^3$ *exponentiate*
$= 7 + 3 \cdot 8$ *multiply*
$= 7 + 24$ *add*
$= 31$

19. Mean $= \dfrac{\text{sum of values}}{\text{number of values}}$

$= \dfrac{4 + 5 + 7 + 9 + 15}{5}$

$= \dfrac{40}{5}$

$= 8$

Because there is an odd number of values, the median is the middle value: Median $= 7$

Range $=$ high $-$ low $= 15 - 4 = 11$

Section 1.1

1. 78: 8 ones, 7 tens

3. 45: 5 ones, 4 tens

5. 348: 8 ones, 4 tens, 3 hundreds

7. 608: 8 ones, 0 tens, 6 hundreds

9. 2,378: 8 ones, 7 tens, 3 hundreds, 2 thousands

11. 273,569: 9 ones, 6 tens, 5 hundreds, 3 thousands, 7 ten thousands, 2 hundred thousands

13. 458,992

ten thousands

15. 507,994,787

↑

hundred millions

17. 267,894,335

↑

ones

19. 4,569,000

hundred thousands

21. $658 = 600 + 50 + 8$

23. $68 = 60 + 8$

25. $4,587 = 4,000 + 500 + 80 + 7$

27. $32,674 = 30,000 + 2,000 + 600 + 70 + 4$

29. $3,462,577 = 3,000,000 + 400,000 + 60,000$
$+ 2,000 + 500 + 70 + 7$

31. $407 = 400 + 7$

33. $30,068 = 30,000 + 60 + 8$

35. $3,004,008 = 3,000,000 + 4,000 + 8$

37. 29 = twenty-nine

39. 40 = forty

41. 573 = five hundred seventy-three

43. 707 = seven hundred seven

45. 770 = seven hundred seventy

47. 23,540 = twenty-three thousand, five hundred forty

49. 3,004 = three thousand, four

51. 3,040 = three thousand, forty

53. 104,065,780 = one hundred four million, sixty-five thousand, seven hundred eighty

55. 5,003,040,008 = five billion, three million, forty thousand, eight

57. 2,546,731 = two million, five hundred forty-six thousand, seven hundred thirty-one

59. Three hundred twenty-five = 325

61. Five thousand, four hundred thirty-two = 5,432

63. Eighty-six thousand, seven hundred sixty-two = 86,762

65. Two million, two hundred = 2,000,200

67. Two million, two thousand, two hundred = 2,002,200

Applying the Concepts

69. 3,100 miles

$\uparrow$

thousands

71. $106,721 = 100,000 + 6,000 + 700 + 20 + 1$

73. 2,555,476 = two million, five hundred fifty-five thousand, four hundred seventy-six

75. Two hundred seventy-five million = 275,000,000

77. 127,000,000 = one hundred twenty-seven million

79. Thirty million = 30,000,000

81. 8,800,000 = eight million, eight hundred thousand

Extending the Concepts

83.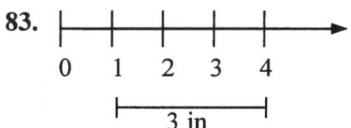

3 inches.

85.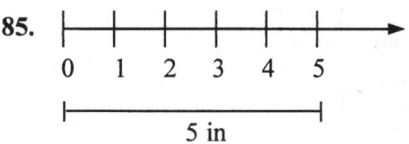

5 inches.

87.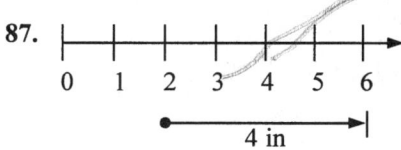

The number 6.

89.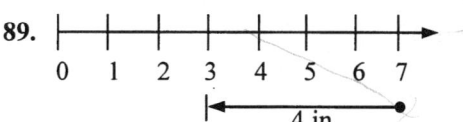

The number 3.

91.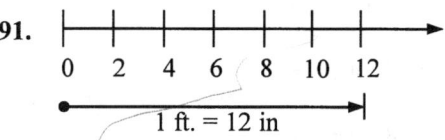

The number 12.

Section 1.2

1. $3 + 5 + 7 = 15$

3. $1 + 4 + 9 = 14$

5. $5 + 9 + 4 + 6 = 24$

7. $1 + 2 + 3 + 4 + 5 = 15$

9. $9 + 1 + 8 + 2 = 20$

11.
$$\begin{array}{r} 43 \\ + 25 \\ \hline 68 \end{array}$$

13.
$$\begin{array}{r} 81 \\ + 17 \\ \hline 98 \end{array}$$

15.
$$\begin{array}{r} 4{,}281 \\ + 3{,}016 \\ \hline 7{,}297 \end{array}$$

17.
$$\begin{array}{r} 3{,}482 \\ + 3{,}005 \\ \hline 6{,}487 \end{array}$$

19.
$$\begin{array}{r} 32 \\ 21 \\ + 43 \\ \hline 96 \end{array}$$

21.
$$\begin{array}{r} 6{,}245 \\ 203 \\ + 1{,}001 \\ \hline 7{,}449 \end{array}$$

23.
$$\begin{array}{r} \overset{1}{4}9 \\ + 16 \\ \hline 65 \end{array}$$

25.
$$\begin{array}{r} \overset{1}{7}4 \\ + 28 \\ \hline 102 \end{array}$$

27.
$$\begin{array}{r} \overset{1}{6}82 \\ + 193 \\ \hline 875 \end{array}$$

29.
$$\begin{array}{r} \overset{1}{6}38 \\ + 191 \\ \hline 829 \end{array}$$

31.
$$\begin{array}{r} \overset{1}{4}{,}\overset{1}{9}63 \\ + 5{,}428 \\ \hline 10{,}391 \end{array}$$

33.
$$\begin{array}{r} 6{,}205 \\ + 9{,}999 \\ \hline 16{,}204 \end{array}$$

35.
$$\begin{array}{r} 56{,}789 \\ + 98{,}765 \\ \hline 155{,}554 \end{array}$$

37.
$$\begin{array}{r} 52{,}468 \\ + 58{,}642 \\ \hline 111{,}110 \end{array}$$

39.
$$\begin{array}{r} {}^{111} \\ 4{,}296 \\ 8{,}720 \\ +\ 4{,}375 \\ \hline 17{,}391 \end{array}$$

41.
$$\begin{array}{r} {}^{112} \\ 4{,}994 \\ 449 \\ +\ 9{,}449 \\ \hline 14{,}892 \end{array}$$

43.
$$\begin{array}{r} {}^{2} \\ 12 \\ 34 \\ 56 \\ +\ 78 \\ \hline 180 \end{array}$$

45.
$$\begin{array}{r} {}^{22} \\ 999 \\ 444 \\ 555 \\ +\ 222 \\ \hline 2{,}220 \end{array}$$

47.
$$\begin{array}{r} {}^{111} \\ 9{,}245 \\ 672 \\ 8{,}341 \\ +\ 27 \\ \hline 18{,}285 \end{array}$$

49.
$$\begin{array}{cccc} 61 & 63 & 65 & 67 \\ +\ 38 & +\ 36 & +\ 34 & +\ 32 \\ \hline 99 & 99 & 99 & 99 \end{array}$$
Note the pattern: Each of the 1st numbers is increased by 2, while each of the 2nd numbers is decreased by 2. The sum remains the same.

51.
$$\begin{array}{cccc} 9 & 36 & 81 & 144 \\ +\ 16 & +\ 64 & +\ 144 & +\ 256 \\ \hline 25 & 100 & 225 & 400 \end{array}$$
Note: All of these numbers are "perfect squares". For example, $9+16=25$ is the same as $3^2+4^2=5^2$.

53. $5+9=9+5$

55. $3+8=8+3$

57. $6+4=4+6$

59. $(1+2)+3=1+(2+3)$

61. $(2+1)+6=2+(1+6)$

63. $1+(9+1)=(1+9)+1$

65. $(4+n)+1=4+(n+1)$

67. The solution to $n+6=10$ is $n=4$, because $4+6=10$.

69. The solution to $n+8=13$ is $n=5$, because $5+8=13$.

71. The solution to $4+n=12$ is $n=8$, because $4+8=12$.

73. The solution to $17=n+9$ is $n=8$, because $17=8+9$.

75. $4\ +\ 9$
$\downarrow$
The sum of 4 and 9.

77. $8\ +\ 1$
$\downarrow$
The sum of 8 and 1.

79.

$$2 + 3 \qquad = \quad 5$$
$$\downarrow \qquad\qquad\quad \downarrow$$
The sum of 2 and 3 is 5.

81. Perimeter = sum of all sides
$$P = 3 \text{ in.} + 3 \text{ in.} + 3 \text{ in.} + 3 \text{ in.}$$
$$P = 12 \text{ in.}$$

83. Perimeter = sum of all sides
$$P = 4 \text{ ft.} + 4 \text{ ft.} + 4 \text{ ft.} + 4 \text{ ft.}$$
$$P = 16 \text{ ft.}$$

85. Perimeter = sum of all sides
$$P = 10 \text{ yd.} + 3 \text{ yd.} + 10 \text{ yd.} + 3 \text{ yd.}$$
$$P = 26 \text{ yd.}$$

87. Perimeter = sum of all sides
$$P = 5 \text{ in.} + 6 \text{ in.} + 7 \text{ in.}$$
$$P = 18 \text{ in.}$$

89.

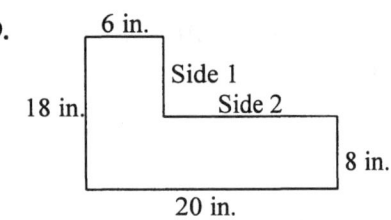

Find lengths of missing sides:
Side 1: 18 in. − 8 in. = 10 in.
Side 2: 20 in. − 6 in. = 14 in.

$$P = 18 \text{ in.} + 6 \text{ in.} + 10 \text{ in.} + 14 \text{ in.} + 8 \text{ in.} + 20 \text{ in.}$$
$$P = 76 \text{ in.}$$

91.

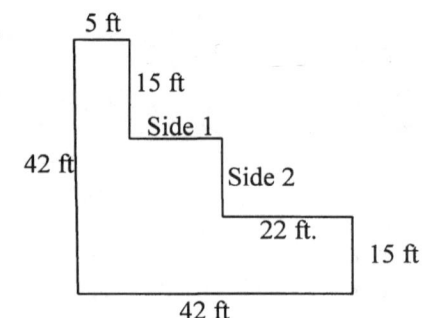

Find lengths of missing sides:
Side 1: 42 ft. − 22 ft. − 5 ft. = 15 ft.
Side 2: 42 ft. − 15 ft. − 15 ft. = 12 ft.
$$P = 42 \text{ ft.} + 5 \text{ ft.} + 15 \text{ ft.} + 15 \text{ ft.} + 12 \text{ ft.} + 22 \text{ ft.}$$
$$+ \; 15 \text{ ft.} + \; 42 \text{ ft.}$$
$$P = 168 \text{ ft.}$$

Applying the Concepts

93.

1st gas purchased:		18 gallons
+ 2nd gas purchased:	+	16 gallons
Total gas purchased:		34 gallons

95.

Monday's balance:		$241
+ Tuesday's deposit:	+	108
Wednesday's balance:		$349

97. a.

Ages 15-19:		425,493
+ Under 15:	+	7,315
Total babies:		432,808 babies

b.

Ages 20-29:		2,082,497
Ages 15-19:		425,493
+ Under 15:	+	7,315
Total babies:		2,515,305 babies

c.

Ages 30-39:		1,405,146
Ages 20-29:		2,082,497
Ages 15-19:		425,493
+ Under 15:	+	7,315
Total babies:		3,920,451 babies

Section 1.3

1. Round 42 to the nearest ten:
locate the digit to the right of the tens:
4<u>2</u>: *2 < 5 so replace it with a 0:*   40

3. Round 46 to the nearest ten:
locate the digit to the right of the tens:
4<u>6</u>: *6 ≥ 5 so replace it with a 0 and add one to the tens place:*   50

5. Round 45 to the nearest ten:
locate the digit to the right of the tens:
4<u>5</u>: *5 ≥ 5 so replace it with a 0 and add one to the tens place:*   50

7. Round 77 to the nearest ten:
locate the digit to the right of the tens:
7<u>7</u>: *7 ≥ 5 so replace it with a 0 and add one to the tens place:*   80

9. Round 458 to the nearest ten:
locate the digit to the right of the tens:
45<u>8</u>: *8 ≥ 5 so replace it with a 0 and add one to the tens place:*   460

11. Round 471 to the nearest ten:
locate the digit to the right of the tens:
47<u>1</u>: *1 < 5 so replace it with a 0:*   470

13. Round 56,782 to the nearest ten:
locate the digit to the right of the tens:
56,78<u>2</u>: *2 < 5 so replace it with a 0:* 56,780

15. Round 4,504 to the nearest ten:
locate the digit to the right of the tens:
4,50<u>4</u>: *4 < 5 so replace it with a 0:*   4,500

17. Round 549 to the nearest hundred:
locate the digit to the right of the hundreds:
5<u>4</u>9: *4 < 5 so replace it and all digits to the right with a 0:*   500

19. Round 833 to the nearest hundred:
locate the digit to the right of the hundreds:
8<u>3</u>3: *3 < 5 so replace it and all digits to the right with a 0:*   800

21. Round 899 to the nearest hundred:
locate the digit to the right of the hundreds:
8<u>9</u>9: *9 ≥ 5 so replace it and all digits to the right with a 0 and add one to the hundreds place.*   900

23. Round 1,090 to the nearest hundred:
locate the digit to the right of the hundreds:
10<u>9</u>0: *9 ≥ 5 so replace it and all digits to the right with a 0 and add one to the hundreds place.*   1,100

25. Round 5,044 to the nearest hundred:
locate the digit to the right of the hundreds:
5,0<u>4</u>4: *4 < 5 so replace it and all digits to the right with a 0:*   5,000

27. Round 39,603 to the nearest hundred:
locate the digit to the right of the hundreds:
39,6<u>0</u>3: *0 < 5 so replace it and all digits to the right with a 0:*   39,600

29. Round 4,670 to the nearest thousand:
locate the digit to the right of the thousands:
4,<u>6</u>70: *6 ≥ 5 so replace it and all digits to the right with a 0 and add one to the thousands place.*   5,000

31. Round 9,760 to the nearest thousand:
locate the digit to the right of the thousands:
9,<u>7</u>60: *7 ≥ 5 so replace it and all digits to the right with a 0 and add one to the thousands place.*   10,000
(Note: you must carry when you add one to the thousands place).

33. Round 978 to the nearest thousand:
locate the digit to the right of the thousands:
<u>9</u>78: *9 ≥ 5 so replace it and all digits to the right with a 0 and add one to the thousands place.*   1,000
(Note: The thousands place was previously 0).

35. Round 657,892 to the nearest thousand:
locate the digit to the right of the thousands:
*657,8**9**2: 8 ≥ 5 so replace it and all digits*
to the right with a 0 and add one to the
thousands place.   658,000

37. Round 509,905 to the nearest thousand:
locate the digit to the right of the thousands:
*509,**9**05: 9 ≥ 5 so replace it and all digits to*
the right with a 0 and add one to the
thousands place.   510,000
(Note: you must carry when you add one to
the thousands place).

39. Round 3,789,345 to the nearest thousand:
locate the digit to the right of the thousands:
*3,789,**3**45: 3 < 5 so replace it and all digits*
to the right with a 0:   3,789,000

41. *Locate the digit to the right of the given*
place value and compare to 5:

Tens	Hundred	Thousand
7,82**1**	7,8**2**1	7,**8**21
1 < 5	2 < 5	8 > 5
7,820	7,800	8,000

43. *Locate the digit to the right of the given*
place value and compare to 5:

Tens	Hundred	Thousand
5,99**9**	5,9**9**9	5,**9**99
9 ≥ 5	9 ≥ 5	9 ≥ 5
6,000	6,000	6,000

Note: Adding one to 9 causes carrying.

45. *Locate the digit to the right of the given*
place value and compare to 5:

Tens	Hundred	Thousand
10,98**5**	10,9**8**5	10,**9**85
5 ≥ 5	8 ≥ 5	9 ≥ 5
10,990	11,000	11,000

Note: Adding one to 9 causes carrying.

47. *Locate the digit to the right of the given*
place value and compare to 5:

Tens	Hundreds	Thousands
99,99**9**	99,9**9**9	99,**9**99
9 ≥ 5	9 ≥ 5	9 ≥ 5
100,000	100,000	100,000

Note: Adding one to 9 causes carrying.

Applying the Concepts

49. Round $2,486,609 to the nearest hundred
thousand:
*2,4**8**6,609: 8 ≥ 5 so replace it and all digits*
to the right with 0 and add 1 to the hundred
thousands place: $2,500,000

51. $7,315 + 425,493 + 2,082,497 + 1,405,146$
$+95,788 + 5,224 + 263$
$= 4,021,726$ babies

53. Round 2,082,497 to the nearest hundred
thousand
*2,0**8**2,497: 8 ≥ 5 so add 1 to the hundred*
thousands place and replace all digits to the
right with 0:   2,100,000 babies

55. From the pie chart:
Supply Expenses $11,456
+ Postage Expenses + 3,792
Postage & Supply Expenses $15,248
To the nearest hundred, $15,248 rounds to
$15,200.

57.
Supply Expenses $11,456
Postage Expenses   3,792
Telephone Expenses   3,652
Car Expenses   3,205
+ All Other Expenses + 8,496
Total $30,601
To the nearest thousand, $30,601 rounds to
$31,000.

Estimating

59.

750	rounds to	800
275	rounds to	300
+ 120	rounds to	+ 100
		1,200

61.

472	rounds to	500
422	rounds to	400
536	rounds to	500
+ 511	rounds to	+ 500
		1,900

63.

25,399	rounds to	25,000
7,601	rounds to	8,000
18,744	rounds to	19,000
+ 6,298	rounds to	+ 6,000
		58,000

65. Draw a horizontal line from the top of the bar above 3 seconds over to the vertical axis. The line is closer to 160 mph.

67. Draw a horizontal line from the top of the bar above 1 second over to the vertical axis. Estimate the speed to be about 75 mph.

69. On the vertical axis, mark the scale at equal intervals of 20, starting at 0 and ending at 100 (the highest level of caffeine). On the horizontal axis, draw a bar of the appropriate height for each drink. Label each bar, and the vertical axis.

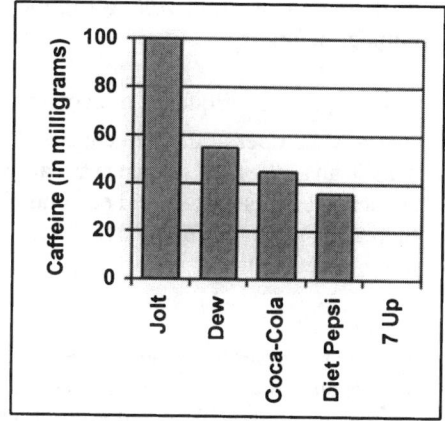

71. On the vertical axis, mark the scale at equal intervals of 100, starting at 0 and ending at 700, (just above the highest level of calories). On the horizontal axis, draw a bar of the appropriate height for each activity. Label each bar, and the vertical axis.

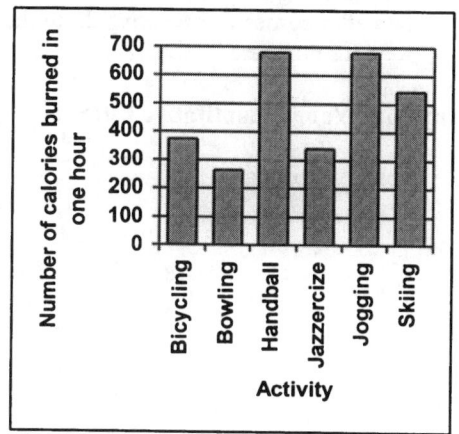

6

Extending the Concepts

73. 1,2,3,4,... is the sequence of counting numbers. The next number in the sequence is 5.
This is an *arithmetic* sequence because the sequence begins with 1, and each number thereafter comes from *adding* 1 to the number before it.

75. 2,4,6,8,... is the sequence of even numbers.
The next number in the sequence is 10.
This is an *arithmetic* sequence because the sequence begins with 2, and each number thereafter comes from *adding* 2 to the number before it.

77. 5,8,11,...
The next number in the sequence is 14.
This is an *arithmetic* sequence because the sequence begins with 5, and each number thereafter comes from *adding* 3 to the previous number.

79. 10,35,60, ...
The next number in the sequence is 85.
This is an *arithmetic* sequence because the sequence begins with 10, and each number thereafter comes from *adding* 25 to the previous number.

Improving Your Quantitative Literacy

81. Answers will vary.

Section 1.4

1. Subtract 24 from 56:
$$\begin{array}{r} 56 \\ -\ 24 \\ \hline 32 \end{array}$$

3. Subtract 23 from 45:
$$\begin{array}{r} 45 \\ -\ 23 \\ \hline 22 \end{array}$$

5. Find the difference of 29 and 19:
$$\begin{array}{r} 29 \\ -\ 19 \\ \hline 10 \end{array}$$

7. Find the difference of 126 and 15:
$$\begin{array}{r} 126 \\ -\ 15 \\ \hline 111 \end{array}$$

9.
$$\begin{array}{r} 975 \\ -\ 663 \\ \hline 312 \end{array}$$

11.
$$\begin{array}{r} 904 \\ -\ 501 \\ \hline 403 \end{array}$$

13.
$$\begin{array}{r} 9{,}876 \\ -\ 8{,}765 \\ \hline 1{,}111 \end{array}$$

15.
$$\begin{array}{r} 7{,}976 \\ -\ 3{,}432 \\ \hline 4{,}544 \end{array}$$

17.
$$\begin{array}{r} {}^{4}\!\!\not{5}\ {}^{12}\!\!\not{2} \\ -\ 3\ 7 \\ \hline 1\ 5 \end{array}$$

19.
$$\begin{array}{r} {}^{6}\!\!\not{7}\ {}^{10}\!\!\not{0} \\ -\ 3\ 7 \\ \hline 3\ 3 \end{array}$$

21.
$$\begin{array}{r} {}^{6}\!\!\not{7}\ {}^{14}\!\!\not{4} \\ -\ 6\ 9 \\ \hline 5 \end{array}$$

23.
$$\begin{array}{r} {}^{4}\!\!\not{5}\ {}^{11}\!\!\not{1} \\ -\ 1\ 8 \\ \hline 3\ 3 \end{array}$$

25.
$$\begin{array}{r} {}^{2}\!\!\not{3}\ {}^{12}\!\!\not{2}\ 9 \\ -\ 2\ 3\ 4 \\ \hline 9\ 5 \end{array}$$

27.
$$\begin{array}{r} {}^{2}\!\!\not{3}\ {}^{14}\!\!\not{4}\ 8 \\ -\ 1\ 9\ 6 \\ \hline 1\ 5\ 2 \end{array}$$

29.
$$\begin{array}{r} {}^{8}\!\!\not{9}\ {}^{12}\!\!\not{3}\ {}^{12}\!\!\not{2} \\ -\ 6\ 5\ 8 \\ \hline 2\ 7\ 4 \end{array}$$

31.
$$\begin{array}{r} {}^{5}\!\!\not{6}\ {}^{13}\!\!\not{4}\ {}^{17}\!\!\not{7} \\ -\ 1\ 5\ 9 \\ \hline 4\ 8\ 8 \end{array}$$

33.
$$\overset{8\ \ 9\ \ 15}{\cancel{9}\,\cancel{0}\,\cancel{5}}$$
$$-\ 3\ 6\ 7$$
$$\overline{\ \ 5\ 3\ 8}$$

35.
$$\overset{5\ \ 9\ \ 10}{\cancel{6}\,\cancel{0}\,\cancel{0}}$$
$$-\ 4\ 3\ 7$$
$$\overline{\ \ 1\ 6\ 3}$$

37.
$$\overset{3\ \ 15}{\cancel{4}\,,\cancel{5}\,8\ 3}$$
$$-\ 2\,,9\ 7\ 3$$
$$\overline{\ \ 1\,,\ 6\ 1\ 0}$$

39.
$$\overset{8\ \ 9\ \ 13\ \ 10}{7\,\cancel{9}\,,\cancel{0}\,\cancel{4}\,\cancel{0}}$$
$$-\ 3\ 2\,,9\ 5\ 7$$
$$\overline{\ \ 4\ 6\,,\ 0\ 8\ 3}$$

41.

$\overset{1\ \ 15}{\cancel{2}\,\cancel{5}}$	$\overset{1\ \ 14}{\cancel{2}\,\cancel{4}}$	$\overset{1\ \ 13}{\cancel{2}\,\cancel{3}}$	$\overset{1\ \ 12}{\cancel{2}\,\cancel{2}}$
$-\ 1\ 5$	$-\ 1\ 6$	$-\ 1\ 7$	$-\ 1\ 8$
$1\ 0$	8	6	4

43.

$\overset{3\ \ 9\ \ 10}{\cancel{4}\,\cancel{0}\,\cancel{0}}$	$\overset{3\ \ 9\ \ 10}{\cancel{4}\,\cancel{0}\,\cancel{0}}$	$\overset{1\ \ 12}{\cancel{2}\,\cancel{2}\,5}$	$\overset{1\ \ 12}{\cancel{2}\,\cancel{2}\,5}$
$-\ 2\ 5\ 6$	$-\ 1\ 4\ 4$	$-\ 1\ 4\ 4$	$-\ \ \ 8\ 1$
$1\ 4\ 4$	$2\ 5\ 6$	$8\ 1$	$1\ 4\ 4$

Applying the Concepts

45.

Opening balance:	$504
− Total checks:	− 249
Amount left in account :	$255

47.

Height in 1954:	29,028 ft.
+ 7 feet:	+ 7 ft.
Height in 1999:	29,035 ft.

Height in 1999:	29,035 ft.
− Height in 1847:	− 29,002 ft.
Difference:	33 ft.

49.

Past enrollment:	567 students
− Current enrollment:	− 399 students
Decrease:	168 students

51.

Opening balance:	$425
+ Deposit:	+ 149
New balance :	$574

53. Total Checks:

Market:	$37
+ Credit Card:	+ 188
Total:	$225

Balance after deposit:	$574
− Total checks:	− 225
New balance :	$349 ≈ $350

55. a. For each year, enter the sales:

Year	Sales (in millions)
1996	386
1997	518
1998	573
1999	819

b.

1999 sales:	$819 million
− 1997 sales:	− 518 million
Increase:	$301 million

Improving Your Quantitative Literacy

57. a. It appears they have been rounded to the nearest hundred.

      b.

2003 participants:	19,100
− 2000 participants:	− 13,000
Increase:	6,100

      c. There was an increase of 6,100 participants from 2000-2003. If the same increase occurs over the next three years:

2003 participants:	19,100
+ increase:	+6,100
2006 participants:	25,200

Section 1.5

1. $3 \cdot 100 = 300$
$3 \cdot 1 = 3$, *attach two 0's*

3. $3 \cdot 200 = 600$
$3 \cdot 2 = 6$, *attach two 0's*

5. $6 \cdot 500 = 3,000$
$6 \cdot 5 = 30$, *attach two 0's*

7. $5 \cdot 1,000 = 5,000$
$5 \cdot 1 = 5$, *attach three 0's*

9. $3 \cdot 7,000 = 21,000$
$3 \cdot 7 = 21$, *attach three 0's*

11. $9 \cdot 9,000 = 81,000$
$9 \cdot 9 = 81$, *attach three 0's*

13.
$$\begin{array}{r} \overset{2}{2}5 \\ \times\ \ 4 \\ \hline 100 \end{array}$$

15.
$$\begin{array}{r} \overset{4}{3}8 \\ \times\ \ 6 \\ \hline 228 \end{array}$$

17.
$$\begin{array}{r} \overset{1}{1}8 \\ \times\ \ 2 \\ \hline 36 \end{array}$$

19.
$$\begin{array}{r} 72 \\ \times\ 20 \\ \hline 00 \\ 1,440 \\ \hline 1,440 \end{array}$$
    $00 \leftarrow 0(72)$
$1,440 \leftarrow 20(72)$
$1,440$ Add

21.
$$\begin{array}{r} 19 \\ \times\ 50 \\ \hline 00 \\ 950 \\ \hline 950 \end{array}$$
    $00 \leftarrow 0(19)$
$950 \leftarrow 50(19)$
950 Add

23.
$$\begin{array}{r} 69 \\ \times\ 25 \\ \hline 345 \\ 1,380 \\ \hline 1,725 \end{array}$$
$345 \leftarrow 5(69)$
$1,380 \leftarrow 20(69)$
$1,725$ Add

25.
$$\begin{array}{r} 11 \\ \times\ 11 \\ \hline 11 \\ 110 \\ \hline 121 \end{array}$$
$11 \leftarrow 1(11)$
$110 \leftarrow 10(11)$
121 Add

27.
$$\begin{array}{r} 97 \\ \times\ 16 \\ \hline 582 \\ 970 \\ \hline 1,552 \end{array}$$
$582 \leftarrow 6(97)$
$970 \leftarrow 10(97)$
$1,552$ Add

29.
$$\begin{array}{r} 168 \\ \times\ 25 \\ \hline 840 \\ 3,360 \\ \hline 4,200 \end{array}$$
$840 \leftarrow 5(168)$
$3,360 \leftarrow 20(168)$
$4,200$ Add

31.
$$\begin{array}{r} 728 \\ \times\ 91 \\ \hline 728 \\ 65,520 \\ \hline 66,248 \end{array}$$
$728 \leftarrow 1(728)$
$65,520 \leftarrow 90(728)$
$66,248$ Add

33.

$$\begin{array}{r} 698 \\ \times\ 400 \\ \hline 279{,}200 \end{array} \leftarrow 400(698)$$

35.

$$\begin{array}{r} 111 \\ \times\ 111 \\ \hline 111 \end{array} \leftarrow 1(111)$$
$$1{,}110 \leftarrow 10(111)$$
$$\underline{11{,}100} \leftarrow 100(111)$$
$$12{,}321 \quad \text{Add}$$

37.

$$\begin{array}{r} 532 \\ \times\ 200 \\ \hline 106{,}400 \end{array} \leftarrow 200(532)$$

39.

$$\begin{array}{r} 856 \\ \times\ 232 \\ \hline 1{,}712 \end{array} \leftarrow 2(856)$$
$$25{,}680 \leftarrow 30(856)$$
$$\underline{171{,}200} \leftarrow 200(856)$$
$$198{,}592 \quad \text{Add}$$

41.

$$\begin{array}{r} 976 \\ \times\ 628 \\ \hline 7{,}808 \end{array} \leftarrow 8(976)$$
$$19{,}520 \leftarrow 20(976)$$
$$\underline{585{,}600} \leftarrow 600(976)$$
$$612{,}928 \quad \text{Add}$$

43.

$$\begin{array}{r} 2{,}468 \\ \times\ 135 \\ \hline 12{,}340 \end{array} \leftarrow 5(2{,}468)$$
$$74{,}040 \leftarrow 30(2{,}468)$$
$$\underline{246{,}800} \leftarrow 100(2{,}468)$$
$$333{,}180 \quad \text{Add}$$

45.

$$\begin{array}{r} 24{,}563 \\ \times\ 735 \\ \hline 122{,}815 \end{array} \leftarrow 5(24{,}563)$$
$$736{,}890 \leftarrow 30(24{,}563)$$
$$\underline{17{,}194{,}100} \leftarrow 700(24{,}563)$$
$$18{,}053{,}805 \quad \text{Add}$$

47.

$$\begin{array}{r} 44{,}777 \\ \times\ 5{,}888 \\ \hline 358{,}216 \end{array} \leftarrow 8(44{,}777)$$
$$3{,}582{,}160 \leftarrow 80(44{,}777)$$
$$35{,}821{,}600 \leftarrow 800(44{,}777)$$
$$\underline{223{,}885{,}000} \leftarrow 5{,}000(44{,}777)$$
$$263{,}646{,}976 \quad \text{Add}$$

49.

11	11	22	22
$\times 11$	$\times 22$	$\times 22$	$\times 44$
11	22	44	88
110	220	440	880
121	242	484	968

Note: In each case, one of the factors is multiplied by 2, and the resulting product is twice the previous product.

51.

25	25	50	50
$\times 15$	$\times 30$	$\times 15$	$\times 30$
125	750	250	1,500
250		500	
375		750	

Note: In each case, one of the factors is multiplied by 2, and the resulting product is twice the previous product.

53.

25	25	25	25
$\times 10$	$\times 100$	$\times 1{,}000$	$\times 10{,}000$
250	2,500	25,000	250,000

Note: When multiplying by a power of 10, add the same number of zeros in the product as are in the power of 10.

55. $6 \cdot 7$
→ The product of 6 and 7

57. $2 \cdot n$
→ The product of 2 and n

59. $9 \cdot 7 = 63$
→ The product of 9 and 7 is 63

61.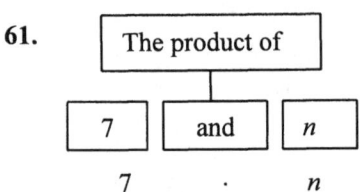

The product of

| 7 | and | n |

$$7 \quad \cdot \quad n$$

63.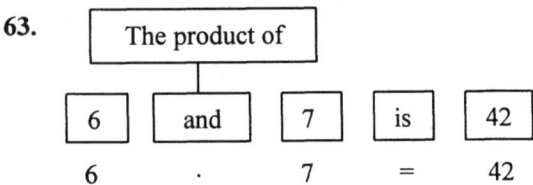

The product of

| 6 | and | 7 | is | 42 |

$$6 \quad \cdot \quad 7 \quad = \quad 42$$

65.

The product of

| 0 | and | 6 | is | 0 |

$$0 \quad \cdot \quad 6 \quad = \quad 0$$

67. Products: $9 \cdot 7, 63$

69. Products: $4(4), 16$

71. Factors: $2, 3, 4$

73. Factors: $2, 2, 3$

75. $5(9) = 9(5)$

77. $6 \cdot 7 = 7 \cdot 6$

79. $2 \cdot (7 \cdot 6) = (2 \cdot 7) \cdot 6$

81. $3 \times (9 \times 1) = (3 \times 9) \times 1$

83. $7(2+3) = 7 \cdot 2 + 7 \cdot 3$ *Distributive Prop.*
           $= 14 + 21$ *Multiply*
           $= 35$ *Add*

85. $9(4+7) = 9 \cdot 4 + 9 \cdot 7$ *Distributive Prop.*
           $= 36 + 63$ *Multiply*
           $= 99$ *Add*

87. $3(x+1) = 3 \cdot x + 3 \cdot 1$ *Distributive Prop.*
           $= 3x + 3$ *Multiply*

89. $2(x+5) = 2 \cdot x + 2 \cdot 5$ *Distributive Prop.*
           $= 2x + 10$ *Multiply*

91. The solution to $4 \cdot n = 12$ is $n = 3$, because $4 \cdot 3 = 12$

93. The solution to $9 \cdot n = 81$ is $n = 9$, because $9 \cdot 9 = 81$

95. The solution to $0 = n \cdot 5$ is $n = 0$, because $0 = 0 \cdot 5$

97. The solution to $27 = 9 \cdot n$ is $n = 3$, because $27 = 9 \cdot 3$

Applying the Concepts

99.

Number of adults:	300
+ Number of children:	+ 80
Number of people attended:	380

101. $300 \text{ adults} \times \dfrac{\$6}{\text{adult ticket}} = \$1,800$

103.

$$\begin{array}{r} \text{Gallons of gas} \quad 130 \\ \underline{\times \text{ miles per gallon} \quad \times \ 22} \\ \text{Total miles} \quad 260 \\ 2{,}600 \\ \underline{2{,}860} \end{array}$$

2,860 miles.

105. 1st 8 hours: $8 \text{ hours} \times \dfrac{\$12}{\text{hour}} = \$96$

Next 4 hours: $4 \text{ hours} \times \dfrac{\$18}{\text{hour}} = \$72$

Total wages: $\$96$

$$\underline{+ \quad 72}$$
$$\$168$$

107. Calories burned playing handball:

Calories per hour: 816

$$\underline{\times \text{ hours} : \quad \times 2}$$

Total calories burned: 1,632

Calories burned bicycling:

Calories per hour: 449

$$\underline{\times \text{ hours} : \quad \times 1}$$

Total calories burned: 449

Handball calories: 1,632

$$\underline{+ \quad \text{Bicycling calories} : \quad + \ 449}$$

Total calories: 2,081

109.

Calories per serving: 140

$$\underline{\times \text{ servings per bag} : \quad \times \ 2}$$

Total calories per bag: 280

111.

Calories per serving: 140

$$\underline{\times \ 3 \text{ servings} : \quad \times 3}$$

Calories per 3 servings: 420

Since you burn 449 calories per hour riding the bike, you would burn off the 420 calories from the chips.

Estimating

113.

750	rounds to	800
$\times$ 12	rounds to	$\times$ 10
		8,000

115.

3,472	rounds to	3,000
$\times$ 511	rounds to	$\times$ 500
		1,500,000

117.

2,399	rounds to	2,000
$\times$ 698	rounds to	$\times$ 700
		1,400,000

Extending the Concepts

119. 5, 10, 20, **40,**…

The geometric sequence starts with 5. Note that $5 \cdot 2 = 10$, $10 \cdot 2 = 20$. Each number thereafter comes from multiplying the previous number by 2. The next number in the sequence is $20 \cdot 2 = 40$.

121. 2, 6, 18, **54,**…

The geometric sequence starts with 2. Note that $2 \cdot 3 = 6$, $6 \cdot 3 = 18$., i.e., each number after 2 in the sequence comes from multiplying the previous number by 3. The next number in the sequence is $18 \cdot 3 = 54$.

Improving Your Quantitative Literacy

123. Each phone pictured in the graph is worth 31 million phones.

a. 31 million phones: 31
   × phones in graph: × 3
   Total phones: 93 million

b. There are between 4 and 5 phones in the graph. Using 5 phones:
   31 million phones: 31
   × phones in graph: × 5
   Total phones: 155 million
   Since there are less than 5 phones in the graph, the statement is false. There will be a less than 155 million.

c. There is between 9 and 10 phones in the graph. Using 10 phones:
   31 million phones: 31
   × phones in graph: × 10
   Total phones: 310 million
   Since there are less than 10 phones in the graph, the statement is true. There will be less than 310 million.

1.

```
        The quotient of
   ┌─────┐  ┌─────┐  ┌─────┐
   │  6  │  │ and │  │  3  │
   └─────┘  └─────┘  └─────┘
      6        ÷        3
```

3.

```
        The quotient of
   ┌─────┐  ┌─────┐  ┌─────┐
   │ 45  │  │ and │  │  9  │
   └─────┘  └─────┘  └─────┘
     45        ÷        9
```

5.

```
        The quotient of
   ┌─────┐  ┌─────┐  ┌─────┐
   │  r  │  │ and │  │  s  │
   └─────┘  └─────┘  └─────┘
      r        ÷        s
```

7.

```
        The quotient of
   ┌─────┐  ┌─────┐  ┌─────┐  ┌─────┐  ┌─────┐
   │ 20  │  │ and │  │  4  │  │ is  │  │  5  │
   └─────┘  └─────┘  └─────┘  └─────┘  └─────┘
     20        ÷        4        =        5
```

9. $6 \div 2 = 3 \Leftrightarrow 2 \cdot 3 = 6$

11. $\dfrac{36}{9} = 4 \Leftrightarrow 9 \cdot 4 = 36$

13. $\dfrac{48}{6} = 8 \Leftrightarrow 6 \cdot 8 = 48$

15. $28 \div 7 = 4 \Leftrightarrow 7 \cdot 4 = 28$

17. $25 \div 5 = 5$, because $5 \cdot 5 = 25$

19. $40 \div 5 = 8$, because $5 \cdot 8 = 40$

21. $9 \div 0$ is undefined. *There is no number that when multiplied by 0 will give you 9.*

23.
$$\begin{array}{r} 45 \\ 8\overline{)360} \\ 32\downarrow \\ \hline 40 \\ -40 \\ \hline 0 \end{array}$$

25.
$$\begin{array}{r} 23 \\ 6\overline{)138} \\ 12\downarrow \\ \hline 18 \\ -18 \\ \hline 0 \end{array}$$

27.
$$\begin{array}{r} 1,530 \\ 5\overline{)7,650} \\ 5\downarrow \\ \hline 26 \\ 25\downarrow \\ \hline 15 \\ 15\downarrow \\ \hline 00 \end{array}$$

29.
$$\begin{array}{r} 1,350 \\ 5\overline{)6,750} \\ 5\downarrow \\ \hline 17 \\ 15\downarrow \\ \hline 25 \\ 25\downarrow \\ \hline 00 \end{array}$$

31.
$$\begin{array}{r} 18,000 \\ 3\overline{)54,000} \\ \underline{3\downarrow} \\ 24 \\ \underline{24\downarrow} \\ 00 \end{array}$$

Note: Fill in the quotient with as many zeros as there are in the dividend.

33.
$$\begin{array}{r} 16,680 \\ 3\overline{)50,040} \\ \underline{3\downarrow} \\ 20 \\ \underline{18\downarrow} \\ 20 \\ \underline{18\downarrow} \\ 24 \\ \underline{24\downarrow} \\ 00 \end{array}$$

Estimating

35. $845 \div 93$ is approximately $900 \div 90 = 10$.
Round the divisor to 90, then round the dividend such that it is easily divisible by 90, such as 900.

37. $15,208 \div 771$ is approximately $16,000 \div 800 = 20$.
Round the divisor to 800, then round the dividend such that it is easily divisible by 800, such as 16,000.

39. $316 \div 289$ is approximately $300 \div 300 = 1$
Round the divisor to 300, then round the dividend to 300 because it is easily divisible by 300.

41. $728 \div 355$ is approximately $800 \div 400 = 2$
Round the divisor to 400, then round the dividend to 800 because it is easily divisible by 400.

43. $921 \div 243$ is approximately $800 \div 200 = 4$
Round the divisor to 200, then round the dividend to 800 because it is easily divisible by 200.

45. $673 \div 109$ is approximately $600 \div 100 = 6$
Round the divisor to 100, then round the dividend to 600 because it is easily divisible by 100.

47.
$$\begin{array}{r} 45 \\ 32\overline{)1,440} \\ \underline{128\downarrow} \\ 160 \\ \underline{160} \\ 0 \end{array}$$

49.
$$\begin{array}{r} 49 \\ 49\overline{)2,401} \\ \underline{196\downarrow} \\ 441 \\ \underline{441} \\ 0 \end{array}$$

51.
$$\begin{array}{r} 432 \\ 28\overline{)12,096} \\ \underline{112\downarrow} \\ 89 \\ \underline{84\downarrow} \\ 56 \\ \underline{56} \\ 0 \end{array}$$

53.
$$
\begin{array}{r}
1,438 \\
63\overline{)90,594} \\
\underline{63\downarrow} \\
275 \\
\underline{252\downarrow} \\
239 \\
\underline{189\downarrow} \\
504 \\
\underline{504} \\
0
\end{array}
$$

55.
$$
\begin{array}{r}
4 \\
25\overline{)100} \\
\underline{100} \\
0
\end{array}
\qquad
\begin{array}{r}
3 \ R22 \\
26\overline{)100} \\
\underline{78} \\
22
\end{array}
$$

$$
\begin{array}{r}
3 \ R19 \\
27\overline{)100} \\
\underline{81} \\
19
\end{array}
\qquad
\begin{array}{r}
3 \ R16 \\
28\overline{)100} \\
\underline{84} \\
16
\end{array}
$$

57.
$$
\begin{array}{r}
61 \ R4 \\
6\overline{)370} \\
\underline{36\downarrow} \\
10 \\
\underline{6} \\
4
\end{array}
$$

59.
$$
\begin{array}{r}
90 \ R1 \\
3\overline{)271} \\
\underline{27\downarrow} \\
01 \\
\underline{0} \\
1
\end{array}
$$

61.
$$
\begin{array}{r}
13 \ R7 \\
26\overline{)345} \\
\underline{26\downarrow} \\
85 \\
\underline{78} \\
7
\end{array}
$$

63.
$$
\begin{array}{r}
234 \ R6 \\
71\overline{)16,620} \\
\underline{142\downarrow} \\
242 \\
\underline{213\downarrow} \\
290 \\
\underline{284} \\
6
\end{array}
$$

65.
$$
\begin{array}{r}
452 \ R4 \\
23\overline{)10,400} \\
\underline{92\downarrow} \\
120 \\
\underline{115\downarrow} \\
50 \\
\underline{46} \\
4
\end{array}
$$

67.
$$
\begin{array}{r}
35 \ R35 \\
169\overline{)5,950} \\
\underline{507\downarrow} \\
880 \\
\underline{845} \\
35
\end{array}
$$

Applying the Concepts

69. $\dfrac{\$22{,}200 \text{ yearly income}}{12 \text{ months per year}} = \text{Monthly income}$

$$\begin{array}{r} 1{,}850 \\ 12\overline{)22{,}200} \\ 12\!\downarrow \\ \hline 102 \\ 96\!\downarrow \\ \hline 60 \\ 60\!\downarrow \\ \hline 00 \end{array}$$

The monthly income is $1,850.

71. $\dfrac{96\cent \text{ (for 6 pounds)}}{6 \text{ pounds}} = \text{cost per pound}$

$$\begin{array}{r} 16 \\ 6\overline{)96} \\ 1\!\downarrow \\ \hline 36 \\ 36 \\ \hline 0 \end{array}$$

One pound costs 16¢.

73. $\dfrac{10{,}000 \text{ steps}}{2{,}000 \text{ steps per mile}} = \text{Number of miles}$

$$\begin{array}{r} 5 \\ 2000\overline{)10000} \\ 10000 \\ \hline 0 \end{array}$$

You need to walk 5 miles.

75. $\dfrac{32 \text{ total ounces}}{5 \text{ ounces per glass}} = \text{Number of glasses}$

$$\begin{array}{r} 6 \\ 5\overline{)32} \\ 30 \\ \hline 2 \end{array}$$

6 glasses with 2 ounces left over.

77.

$$\begin{array}{lr} 16 \text{ glasses} & 16 \\ \times\, 6 \text{ ounces per glass:} & \times\, 6 \\ \hline \text{Total ounces needed:} & 96 \end{array}$$

$\dfrac{96 \text{ total ounces needed}}{32 \text{ ounces per bottle}} = \text{Number of bottles}$

$$\begin{array}{r} 3 \\ 32\overline{)96} \\ 96 \\ \hline 0 \end{array}$$

You will need 3 bottles.

79. $\dfrac{\$192 \text{ (for 16 bottles)}}{16 \text{ bottles}} = \text{Cost per bottle}$

$$\begin{array}{r} 12 \\ 16\overline{)192} \\ 16\!\downarrow \\ \hline 32 \\ 32 \\ \hline 0 \end{array}$$

$12 per bottle.

81. Eggs: 25 mg per egg · 3 eggs = 75 mg
Milk: 215 mg per glass · 2 glasses = 430 mg
Toast: 40 mg per slice · 4 slices = 160 mg

$$\begin{array}{lr} \text{Calcium in eggs:} & 75 \text{ mg} \\ \text{Calcium in milk:} & 430 \text{ mg} \\ +\,\text{Calcium in toast:} & +\,160 \text{ mg} \\ \hline \text{Calcium in breakfast:} & 665 \text{ mg} \end{array}$$

Calculator Problems

83. $305{,}026 \div 698 = 437$

85. $18{,}436{,}466 \div 5{,}678 = 3{,}247$

87. $695 \overline{)603{,}955} = 603{,}955 \div 695 = 869$

89. $4{,}903 \overline{)27{,}868{,}652} = 27{,}868{,}652 \div 4{,}903$
$$= 5{,}684$$

91. $\dfrac{79{,}768 \text{ total gallons}}{472 \text{ minutes}} = \text{Gallons per minute}$

$79{,}768 \div 472 = 169$
169 gallons each minute.

Section 1.7

1. 4^5: base = 4, exponent = 5

3. 3^6: base = 3, exponent = 6

5. 8^2: base = 8, exponent = 2

7. 9^1: base = 9, exponent = 1

9. 4^0: base = 4, exponent = 0

11. $6^2 = 6 \cdot 6 = 36$

13. $2^3 = 2 \cdot 2 \cdot 2 = 8$

15. $1^4 = 1 \cdot 1 \cdot 1 \cdot 1 = 1$

17. $9^0 = 1$ *by definition:* $a^0 = 1$

19. $9^2 = 9 \cdot 9 = 81$

21. $10^1 = 10$

23. $12^1 = 12$

25. $45^0 = 1$ *by definition:* $a^0 = 1$

27. $3 + 5 \cdot 8$ *multiply*
$= 3 + 40$ *add*
$= 43$

29. $3 \cdot 6 - 2$ *multiply*
$= 18 - 2$ *subtract*
$= 16$

31. $6 \cdot 2 + 9 \cdot 8$ *multiply 1st*
$= 12 + 72$ *add*
$= 84$

33. $8 \cdot 10^2 - 6 \cdot 4^3$ *exponentiate*
$= 8 \cdot 100 - 6 \cdot 64$ *multiply*
$= 800 - 384$ *subtract*
$= 416$

35. $2(3 + 6 \cdot 5)$ *inside parenthesis*
$= 2(3 + 30)$ *inside parenthesis*
$= 2(33)$ *multiply*
$= 66$

37. $19 + 50 \div 5^2$ *exponentiate*
$= 19 + 50 \div 25$ *divide*
$= 19 + 2$ *add*
$= 21$

39. $9 - 2(4 - 3)$ *parenthesis*
$= 9 - 2(1)$ *multiply*
$= 9 - 2$ *subtract*
$= 7$

41. $8 \cdot 2^4 + 25 \div 5 - 3^2$ *exponentiate*
$= 8 \cdot 16 + 25 \div 5 - 9$ *multiply & divide*
$= 128 + 5 - 9$ *add & subtract*
$= 124$

43. $5 + 2\left[9 - 2(4 - 1)\right]$ *inside parenthesis*
$= 5 + 2\left[9 - 2 \cdot 3\right]$ *inside brackets*
$= 5 + 2\left[9 - 6\right]$ *inside brackets*
$= 5 + 2 \cdot 3$ *multiply*
$= 5 + 6$ *add*
$= 11$

45. $3+4[6+8(2-0)]$ *inside parenthesis*

$=3+4[6+8\cdot2]$ *inside brackets*

$=3+4[6+16]$ *inside brackets*

$=3+4\cdot22$ *multiply*

$=3+88$ *add*

$=91$

47.

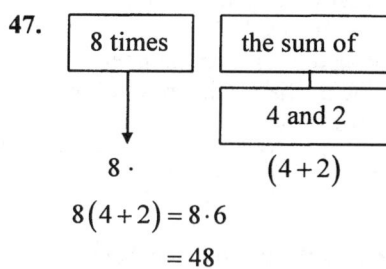

$8(4+2)=8\cdot6$

$=48$

49.

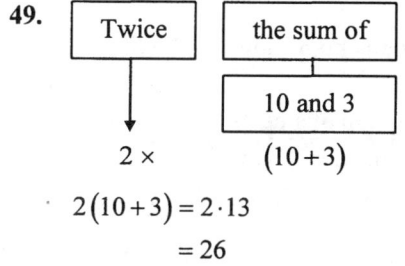

$2(10+3)=2\cdot13$

$=26$

51.

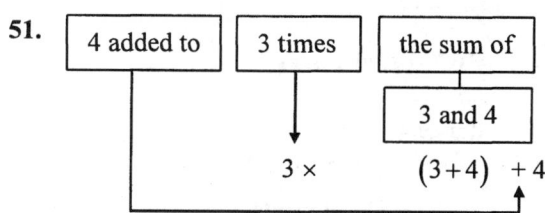

$3(3+4)+4=3\cdot7+4$

$=21+4$

$=25$

53.

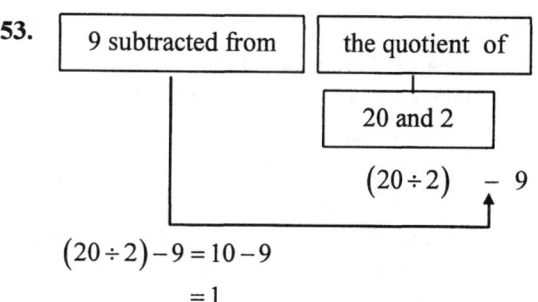

$(20\div2)-9=10-9$

$=1$

55.

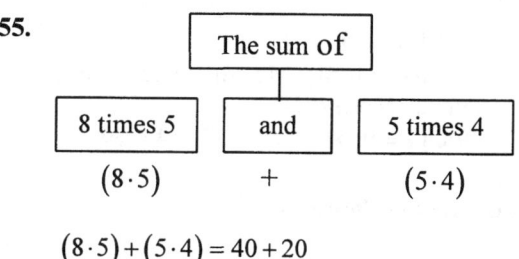

$(8\cdot5)+(5\cdot4)=40+20$

$=60$

57. $\text{Mean}=\dfrac{\text{sum of values}}{\text{number of values}}$

$=\dfrac{1+2+3+4+5}{5}$

$=\dfrac{15}{5}$

$=3$

Range $=$ high value $-$ low value

$=5-1$

$=4$

59. $\text{Mean}=\dfrac{\text{sum of values}}{\text{number of values}}$

$=\dfrac{1+3+9+11}{4}$

$=\dfrac{24}{4}$

$=6$

Range $=$ high value $-$ low value

$=11-1$

$=10$

61. $5, 9, 11, 13, 15$
Because there is an odd number of values,
the median is the middle value: Median $= 11$
Range $=$ high $-$ low $= 15 - 5 = 10$

63. $10, 20, 50, 90, 100$
Because there is an odd number of values,
the median is the middle value: Median $= 50$
Range $=$ high $-$ low $= 100 - 10 = 90$

65. $14, 18, 27, 36, 18, 73$
The mode is the value that occurs most
often. Mode $= 18$.
Range $=$ high $-$ low $= 73 - 14 = 59$

Applying the Concepts

67. Calories in:

1 serving of spaghetti:	210
1 serving of tomatoes:	25
+ 1 serving of cheese:	+ 20
Total calories:	255

69. Calories in:

2 servings of spaghetti:	420
1 serving of tomatoes:	25
+ 1 serving of cheese:	+ 20
Total calories:	465

71. Fat calories in:

2 serving of spaghetti:	20
1 serving of tomatoes:	0
+ 1 serving of cheese:	+ 10
Total calories from fat:	30

73.

Big Mac calories:	510
$-$ Meal calories:	$-$ 255
Difference in calories:	255

The Big Mac has 255 more calories, or
twice as many calories, as the meal.

75. a. Mean $= \dfrac{\text{sum of points}}{\text{number of games}}$

$$= \frac{61 + 76 + 98 + 55 + 76 + 102}{6}$$

$$= \frac{468}{6}$$

$$= 78 \text{ points}$$

b. Put in order from smallest to largest:
55, 61 ,76, 76, 98, 102
Because there is an even number of
values, the median is the mean of the
middle two values:

$$\text{Median} = \frac{76 + 76}{2} = \frac{152}{2} = 76$$

c. The mode is the value that occurs most
often: mode $= 76$

d. Range $=$ high $-$ low $= 102 - 55 = 47$

77. Mean $= \dfrac{\text{sum of stuents}}{\text{number of years}}$

$$= \frac{6,789 + 6,970 + 7,242 + 6,981 + 6,423}{5}$$

$$= \frac{34,405}{5}$$

$$= 6,881 \text{ students}$$

Range $=$ high $-$ low
$= 7,242 - 6,423$
$= 819$ students

Extending the Concepts

79. $1, 3, 5, 7...$
Add the 1st 2 numbers: $1 + 3 = 4$

81. $1, 3, 5, 7...$
Add the 1st 4 numbers: $1 + 3 + 5 + 7 = 16$

83. Answers will vary.

85. 1, 1, 2, 3, 5, 8, …

The first term in the sequence is 1. The second term in the sequence is 1. Thereafter, each term in the sequence is obtained by taking the sum of the 2 previous terms, i.e.,

$1 + 1 = 2$

$1 + 2 = 3$

$2 + 3 = 5$

$3 + 5 = 8$

$5 + 8 = 13$

$8 + 13 = 21$

$13 + 21 = 34$

The next three numbers in the sequence are 13, 21, 34.

Section 1.8

1. Square:

$$A = s^2$$
$$A = (5 \text{ cm})^2$$
$$A = 25 \text{ cm}^2$$

3. Rectangle:

$$A = l \cdot w$$
$$A = (24 \text{ m})(14 \text{ m})$$
$$A = 336 \text{ m}^2$$

5. Parallelogram:

$$A = b \cdot h$$
$$A = (10 \text{ ft})(6 \text{ ft})$$
$$A = 60 \text{ ft}^2$$

7.

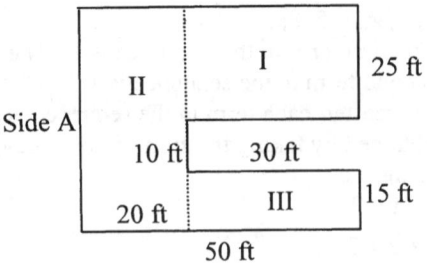

Separate into 3 areas.
Find missing dimension:

Side A = 15 ft. + 10 ft. + 25 ft. = 50 ft.

Area I = lw

$$= (25 \text{ ft})(30 \text{ ft})$$
$$= 750 \text{ ft}^2$$

Area II = lw

$$= (50 \text{ ft})(20 \text{ ft})$$
$$= 1,000 \text{ ft}^2$$

Area III = lw

$$= (30 \text{ ft})(15 \text{ ft})$$
$$= 450 \text{ ft}^2$$

Area = Area I + Area II + Area III

$$= 750 \text{ ft}^2 + 1,000 \text{ ft}^2 + 450 \text{ ft}^2$$
$$= 2,200 \text{ ft}^2$$

9.

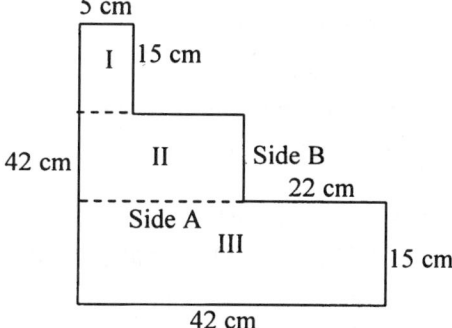

Separate into 3 areas.
Find missing dimension:
Side A = 42 cm − 22 cm = 20 cm
Side B =
42 cm − 15 cm − 15 cm = 12 cm
Area I = lw

$= (5\ \text{cm})(15\ \text{cm})$

$= 75\ \text{cm}^2$

Area II = lw

$= (20\ \text{cm})(12\ \text{cm})$

$= 240\ \text{cm}^2$

Area III = lw

$= (42\ \text{cm})(15\ \text{cm})$

$= 630\ \text{cm}^2$

Area = Area I + Area II + Area III

$= 75\ \text{cm}^2 + 240\ \text{cm}^2 + 630\ \text{cm}^2$

$= 945\ \text{cm}^2$

11. Square:

$A = s^2$

$A = (10\ \text{in})^2$

$A = 100\ \text{in}^2$

13. Cube:

$V = s^3$

$V = (4\ \text{cm})^3$

$V = 64\ \text{cm}^3$

15.

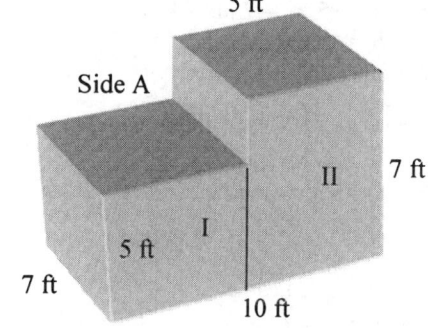

Separate into 2 rectangular solids.
Find missing dimension:
Side A = 10 ft − 5 ft = 5 ft
Volume I = $l \cdot w \cdot h$

$= (5\ \text{ft})(7\ \text{ft})(5\ \text{ft})$

$= 175\ \text{ft}^3$

Volume II = $l \cdot w \cdot h$

$= (5\ \text{ft})(7\ \text{ft})(7\ \text{ft})$

$= 245\ \text{ft}^3$

Volume = Volume I + Volume II

$= 175\ \text{ft}^3 + 420\ \text{ft}^3$

$= 420\ \text{ft}^3$

17.

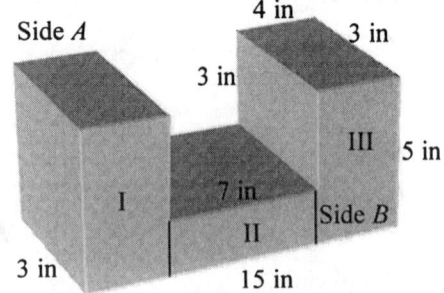

Side A

4 in

3 in

3 in

III 5 in

I

7 in

Side B

II

3 in

15 in

Separate into 3 rectangular solids.
Find in missing dimensions:
Side A = 15 in. − 7 in. − 4 in. = 4 in.
Side B = 5 in. − 3 in. = 2 in.
Volume I = $l \cdot w \cdot h$

$$= (3 \text{ in})(4 \text{ in})(5 \text{ in})$$

$$= 60 \text{ in}^3$$

Volume II = $l \cdot w \cdot h$

$$= (3 \text{ in})(7 \text{ in})(2 \text{ in})$$

$$= 42 \text{ in}^3$$

Volume III = $l \cdot w \cdot h$

$$= (4 \text{ in})(5 \text{ in})(3 \text{ in})$$

$$= 60 \text{ in}^3$$

Volume = Vol I + Vol II + Vol III

$$= 60 \text{ in}^3 + 42 \text{ in}^3 + 60 \text{ in}^3$$

$$= 162 \text{ in}^3$$

19. $S = 2lw + 2lh + 2hw$

$$= 2(4)(4) + 2(4)(4) + 2(4)(4)$$

$$= 32 \text{ cm}^2 + 32 \text{ cm}^2 + 32 \text{ cm}^2$$

$$= 96 \text{ cm}^2$$

21. $S = 2lw + 2lh + 2hw$

$$= 2 \cdot (5\text{ft})(4\text{ft}) + 2(5\text{ft})(6\text{ft}) + 2(6\text{ft})(4\text{ft})$$

$$= 40 \text{ ft}^2 + 60 \text{ ft}^2 + 48 \text{ ft}^2$$

$$= 148 \text{ ft}^2$$

23. $S = 2lw + 2lh + 2hw$

$$= 2(10)(8) + 2(10)(8) + 2(8)(8)$$

$$= 160 \text{ ft}^2 + 160 \text{ ft}^2 + 128 \text{ ft}^2$$

$$= 448 \text{ ft}^2$$

Applying the Concepts

25.

20 ft

40 ft

Make a sketch. Find the perimeter of the pool:

$$P = 2l + 2w$$

$$P = 2 \cdot 40 \text{ ft} + 2 \cdot 20 \text{ ft}$$

$$P = 80 \text{ ft} + 40 \text{ ft}$$

$$P = 120 \text{ ft}$$

Since each tile is 1 foot square, 120 tiles will cover the *inside* perimeter. However, from the diagram you can see you need *4 additional tiles* for the corners. Total tiles: 124

27.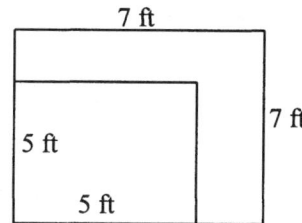

Make a sketch. Find the area of the large and small squares:

Small square: Large square:

$A = s^2$ $A = s^2$

$A = (5 \text{ ft})^2$ $A = (7 \text{ ft})$

$A = 25 \text{ ft}^2$ $A = 49 \text{ ft}^2$

Increase in area $= 49 \text{ ft}^2 - 25 \text{ ft}^2 = 24 \text{ ft}^2$

29. Rectangle:

$A = l \cdot w$

$A = (127 \text{ mm})(67 \text{ mm})$

$A = 8,509 \text{ mm}^2$

31. Rectangle:

$A = l \cdot w$

$A = (36 \text{ mm})(20 \text{ mm})$

$A = 720 \text{ mm}^2$

33. $V = l \cdot w \cdot h$

$\quad = (13 \text{ ft})(13 \text{ ft})(8 \text{ ft})$

$\quad = 1,352 \text{ ft}^3$

35. a. The squares are as follows:

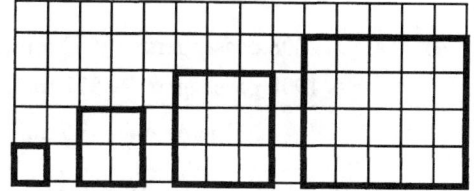

b. The tables are as follows:

Length of side, s (in cm)	$P = 4s$ (in cm)
1	$P = 4 \cdot 1 = 4$
2	$P = 4 \cdot 2 = 8$
3	$P = 4 \cdot 3 = 12$
4	$P = 4 \cdot 4 = 16$

Length of side, s (in cm)	$A = s^2$ (in cm^2)
1	$P = 1^2 = 1$
2	$P = 2^2 = 4$
3	$P = 3^2 = 9$
4	$P = 4^2 = 16$

Extending the Concepts

37. $\quad A = s^2$

$\quad 49 \text{ ft}^2 = s^2$

$\quad 49 \text{ ft}^2 = (7 \text{ ft})^2$ *by inspection, $s = 7$ ft*

Each side is 7 ft.

39. $\quad A = l \cdot w$

$\quad 36 \text{ ft}^2 = l \cdot 4 \text{ ft}$

$\quad 36 \text{ ft}^2 = 9 \text{ ft} \cdot 4 \text{ ft}$ *by inspection, $l = 9$ ft*

The length is 9 ft.

Improving Your Quantitative Literacy

41. a.
$$
\begin{array}{rl}
\text{2008 passengers:} & \text{653 million} \\
-\text{ 1998 passengers:} & -\text{ 577 million} \\
\hline
\text{Increase:} & \text{76 million}
\end{array}
$$

b. Since there were 615 million passengers in 2000, that number will next occur in 2007 (since 2006 has 610 million, and 2008 has 653 million).

c. Answers will vary.

Chapter 1 Review

1. 1,376 = one thousand, three hundred
seventy-six

3. Five million, two hundred forty-five
thousand, six hundred fifty-two = 5,245,652

5. $1,025,639 = 1,000,000 + 20,000 + 5,000$
$+ 600 + 30 + 9$

7. $5 + 7 = 7 + 5$
d, Commutative Property of Addition

9. $6 \cdot 1 = 6$
c, Multiplicative Property of 1

11. $5 \cdot 0 = 0$
b, Multiplicative Property of 0

13. $5 \cdot (3 \cdot 2) = (5 \cdot 3) \cdot 2$
g, Associative Property of Multiplication

15.
$$\begin{array}{r} \overset{1}{4}98 \\ +\ 251 \\ \hline 749 \end{array}$$

17.
$$\begin{array}{r} \overset{1\ \ 11}{7,384} \\ 251 \\ +\ 637 \\ \hline 8,272 \end{array}$$

19.
$$\begin{array}{r} 789 \\ -\ 475 \\ \hline 314 \end{array}$$

21.
$$\begin{array}{r} \overset{8\ \ 9\ \ 18}{5,\cancel{8}\cancel{8}\cancel{8}} \\ -2,7\ 5\ 9 \\ \hline 3,1\ 4\ 9 \end{array}$$

23.
$$\begin{array}{r} \overset{2}{7}3 \\ \times\ 8 \\ \hline 584 \end{array}$$

25.
$$\begin{array}{r} 63 \\ \times 59 \\ \hline 567 \\ 3,150 \\ \hline 3,717 \end{array}$$

27.
$$\begin{array}{r} 173 \\ 4\overline{)692} \\ 4\downarrow \\ \hline 29 \\ 28\downarrow \\ \hline 12 \\ 12 \\ \hline 0 \end{array}$$

29.
$$\begin{array}{r} 428 \\ 36\overline{)15,408} \\ 144\downarrow \\ \hline 100 \\ -72\downarrow \\ \hline 288 \\ 288 \\ \hline 0 \end{array}$$

31. Round 3,781,092 to the nearest ten:
locate the digit to the right of the tens:
3,781,09**2**: *2 < 5 so replace it with 0 :*
3,781,090

33. Round 3,781,092 to the nearest hundred thousand:
locate the digit to the right of the hundred thousands:
3,7**8**1,092: $8 \geq 5$ *so replace it and all digits to the right with 0 and add 1 to the hundred thousands place:*

$$3,800,000$$

35. $4 + 3 \cdot 5^2$ *exponentiate*
$= 4 + 3 \cdot 25$ *multiply*
$= 4 + 75$ *add*
$= 79$

37. $3(2 + 8 \cdot 9)$ *parenthesis (multiply)*
$= 3(2 + 72)$ *parenthesis (add)*
$= 3(74)$ *multiply*
$= 222$

39. $24 \div 6 \cdot 2$ *divide*
$= 4 \cdot 2$ *multiply*
$= 8$

41. $4(3 - 1)^3$ *parenthesis*
$= 4(2)^3$ *exponentiate*
$= 4(8)$ *multiply*
$= 32$

43. $\text{Mean} = \dfrac{\text{sum of values}}{\text{number of values}}$
$= \dfrac{80 + 67 + 78 + 91}{4}$
$= \dfrac{316}{4}$
$= 79$

Put the grades in order: 67, 78, 80, 91
Since there are an even number of values, the median is the mean of the middle two:
$$\text{Median} = \dfrac{78 + 80}{2} = \dfrac{158}{2} = 79$$

45.

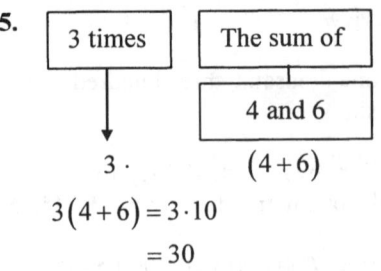

$$3(4+6) = 3 \cdot 10$$
$$= 30$$

47.

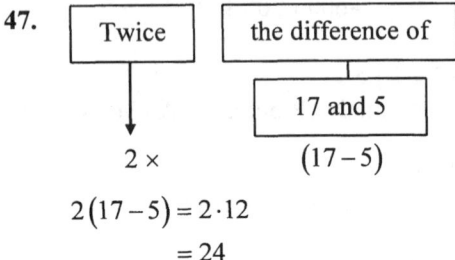

$$2(17 - 5) = 2 \cdot 12$$
$$= 24$$

Applying the Concepts

49.
Income :	$1,783
− Expenses :	− 1,295
Difference:	$488

51. Smallest soccer field:
$P = 2l + 2w$
$= 2(100 \text{ yd}) + 2(55 \text{ yd})$
$= 200 \text{ yd} + 110 \text{ yd}$
$= 310 \text{ yd}$
Rose Bowl:
$P = 2l + 2w$
$= 2(116 \text{ yd}) + 2(72 \text{ yd})$
$= 232 \text{ yd} + 144 \text{ yd}$
$= 376 \text{ yd}$
Largest soccer field:
$P = 2l + 2w$
$= 2(120 \text{ yd}) + 2(75 \text{ yd})$
$= 240 \text{ yd} + 150 \text{ yd}$
$= 390 \text{ yd}$

53.

	Rent:	$1,150
	Food:	625
+	Entertainment:	+ 257
	Total:	$2,032

55. $\dfrac{\$23,256 \text{ yearly income}}{12 \text{ months per year}} = \text{Monthly income}$

$$
\begin{array}{r}
1,938 \\
12\overline{)23,256} \\
\underline{12\downarrow} \\
112 \\
\underline{108\downarrow} \\
45 \\
\underline{36\downarrow} \\
96 \\
\underline{96} \\
0
\end{array}
$$

The monthly income is $1,938.

57. 1st 40 hours: $40 \text{ hours} \times \dfrac{\$16}{\text{hour}} = \$640$

Overtime: $5 \text{ hours} \times \dfrac{\$24}{\text{hour}} = \$120$

Gross pay: $640

$ \underline{+ \ 120}$

$ \$760$

Take-home pay:	Gross pay:	$760
	– deductions:	– 228
	Take-home pay:	$532

59.

Calories in Big Mac:	510
× 2:	×2
Calories in 2 Big Macs:	1,020

Calories in:

2 Big Macs:	1,020
+ 1 order fries:	+ 450
Total calories:	1,470

61.

Calories in burrito:	345
× 2:	×2
Calories in 2 burritos:	690

Calories in:

1 Colossus burger:	940
– 2 burritos:	– 690
Difference:	250

There are 250 more calories.

63. Calories burned in 30 minutes: 235

Calories burned in 60 minutes:

$235 \cdot 2 = 470$

Since a Whopper contains 630 calories, you would not burn enough calories skating.

65. Calories in:

1 Big Mac:	510
+ 1 large fries:	+ 450
Total calories:	960

You could choose many combinations of activities to total at least 960 calories.

67. $V = l \cdot w \cdot h$

$ = (8 \text{ cm})(4 \text{ cm})(5 \text{ cm})$

$ = 160 \text{ cm}^3$

$S = 2lw + 2lh + 2hw$

$ = 2(8)(4) + 2(8)(5) + 2(5)(4)$

$ = 64 \text{ cm}^2 + 80 \text{ cm}^2 + 40 \text{ cm}^2$

$ = 184 \text{ cm}^2$

Chapter 1 Test

1. 20,347 = Twenty thousand, three hundred forty-seven.

3. $123,407 = 100,000 + 20,000 + 3,000$
$$+ 400 + 7$$

5. $7 \cdot 1 = 7$
c, Multiplicative Property of 1

7. $5 \cdot 6 = 6 \cdot 5$
Commutative Property of Multiplication

9.
$$\begin{array}{r} \overset{1}{5},\overset{1}{4}01 \\ 329 \\ +\ 10,653 \\ \hline 16,383 \end{array}$$

11.
$$\begin{array}{r} \overset{6}{\cancel{7}},\overset{9}{\cancel{0}}\overset{14}{\cancel{5}}\overset{12}{\cancel{2}} \\ -\ 3,967 \\ \hline 3,085 \end{array}$$

13.
$$\begin{array}{r} 359 \\ \times\ 62 \\ \hline 718 \\ 21,540 \\ \hline 22,258 \end{array}$$

15.
$$\begin{array}{r} 21 \\ 583\overline{)12243} \\ 1166\downarrow \\ \hline 583 \\ \underline{583} \\ 0 \end{array}$$

17. $8(5)^2 - 7(3)^3$ *exponentiate*
$= 8 \cdot 25 - 7 \cdot 27$ *multiply*
$= 200 - 189$ *subtract*
$= 11$

19. $7 + 2(53 - 3)$ *parenthesis*
$= 7 + 2(50)$ *multiply*
$= 7 + 100$ *add*
$= 107$

21. Sum of prices = $210,000 + $139,000 + $122,000 + $145,000 + $120,000 + $540,000 + $167,000 + $125,000 + $125,000 + $950,000 = $2,643,000

$$\text{Mean} = \frac{\text{sum of values}}{\text{number of values}}$$
$$= \frac{\$2,643,000}{10}$$
$$= \$264,300$$

Median: Order the prices from low to high:
$120,000, $122,000, $125,000, $125000, $139,000, $145,000, $167,000, $210,000, $540,000, $950,000

Since there are an even number of values, the median is the mean of the middle two:
$$\text{Median} = \frac{\$139,000 + \$145,000}{2}$$
$$= \frac{\$284,000}{2}$$
$$= \$142,000$$

Mode: occurs most often: $125,000

23.

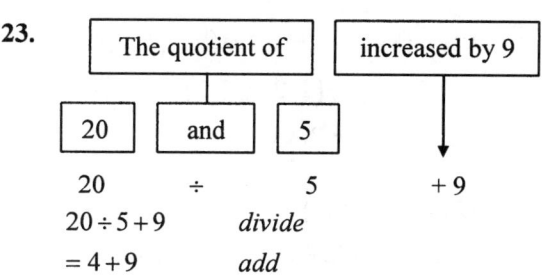

$$20 \qquad \div \qquad 5 \qquad +9$$

$20 \div 5 + 9 \qquad$ *divide*

$= 4 + 9 \qquad$ *add*

$= 13$

25. $\quad P = 2l + 2w$

$\qquad = 2(4 \text{ ft}) + 2(3 \text{ ft})$

$\qquad = 8 \text{ ft} + 6 \text{ ft}$

$\qquad = 14 \text{ ft}$

$\quad A = lw$

$\qquad = (4 \text{ ft})(3 \text{ ft})$

$\qquad = 12 \text{ ft}^2$

Chapter 1: A Glimpse of Algebra

1. $(6x)^2 = (6x)(6x)$
$\qquad = (6\cdot 6)(x\cdot x)$
$\qquad = 36x^2$

3. $(4x)^2 = (4x)(4x)$
$\qquad = (4\cdot 4)(x\cdot x)$
$\qquad = 16x^2$

5. $(3a)^3 = (3a)(3a)(3a)$
$\qquad = (3\cdot 3\cdot 3)(a\cdot a\cdot a)$
$\qquad = 27a^3$

7. $(2ab)^3 = (2ab)(2ab)(2ab)$
$\qquad = (2\cdot 2\cdot 2)(a\cdot a\cdot a)(b\cdot b\cdot b)$
$\qquad = 8a^3b^3$

9. $(9xy)^2 = (9xy)(9xy)$
$\qquad = (9\cdot 9)(x\cdot x)(y\cdot y)$
$\qquad = 81x^2y^2$

11. $(5xyz)^2 = (5xyz)(5xyz)$
$\qquad = (5\cdot 5)(x\cdot x)(y\cdot y)(z\cdot z)$
$\qquad = 25x^2y^2z^2$

13. $(4x)^2(9xy)^2 = (4x)(4x)(9xy)(9xy)$
$\qquad = (4\cdot 4\cdot 9\cdot 9)(x\cdot x\cdot x\cdot x)(y\cdot y)$
$\qquad = 1{,}296x^4y^2$

15. $(2x)^2(3x)^2(4x)^2$
$\qquad = (2x)(2x)(3x)(3x)(4x)(4x)$
$\qquad = (2\cdot 2\cdot 3\cdot 3\cdot 4\cdot 4)(x\cdot x\cdot x\cdot x\cdot x\cdot x)$
$\qquad = 576x^6$

17. $(2x)^3(5x)^2 = (2x)(2x)(2x)(5x)(5x)$
$\qquad = (2\cdot 2\cdot 2\cdot 5\cdot 5)(x\cdot x\cdot x\cdot x\cdot x)$
$\qquad = 200x^5$

19. $(2a)^3(3a)^2(10a)^2$
$\qquad = (2a)(2a)(2a)(3a)(3a)(10a)(10a)$
$\qquad = (2\cdot 2\cdot 2\cdot 3\cdot 3\cdot 10\cdot 10)(a\cdot a\cdot a\cdot a\cdot a\cdot a\cdot a)$
$\qquad = 7{,}200a^7$

21. $(3xy)^3(4xy)^2$
$\qquad = (3xy)(3xy)(3xy)(4xy)(4xy)$
$\qquad = (3\cdot 3\cdot 3\cdot 4\cdot 4)(x\cdot x\cdot x\cdot x\cdot x)(y\cdot y\cdot y\cdot y\cdot y)$
$\qquad = 432x^5y^5$

23. $(5xyz)^2(2xyz)^4$
$\qquad = (5xyz)(5xyz)(2xyz)(2xyz)(2xyz)(2xyz)$
$\qquad = (5\cdot 5\cdot 2\cdot 2\cdot 2\cdot 2)(x\cdot x\cdot x\cdot x\cdot x\cdot x)$
$\qquad \times (y\cdot y\cdot y\cdot y\cdot y\cdot y)(z\cdot z\cdot z\cdot z\cdot z\cdot z)$
$\qquad = 400x^6y^6z^6$

25. $(xy)^3(xz)^2(yz)^4$
$\qquad = (xy)(xy)(xy)(xz)(xz)(yz)(yz)(yz)(yz)$
$\qquad = (x\cdot x\cdot x\cdot x\cdot x)(y\cdot y\cdot y\cdot y\cdot y\cdot y\cdot y)$
$\qquad \times (z\cdot z\cdot z\cdot z\cdot z\cdot z)$
$\qquad = x^5y^7z^6$

27. $(2a^3b^2)^2(3a^2b^3)^4$
$\qquad = (2a^3b^2)(2a^3b^2)(3a^2b^3)(3a^2b^3)(3a^2b^3)$
$\qquad \times (3a^2b^3)$
$\qquad = (2\cdot 2\cdot 3\cdot 3\cdot 3\cdot 3)(a\cdot a\cdot a\cdot a\cdot a\cdot a\cdot a\cdot a\cdot a\cdot a\cdot a\cdot a\cdot a\cdot a)$
$\qquad \times (b\cdot b\cdot b\cdot b\cdot b\cdot b\cdot b\cdot b\cdot b\cdot b\cdot b\cdot b\cdot b\cdot b\cdot b\cdot b)$
$\qquad = 324a^{14}b^{16}$

29. $\left(5x^2y^3\right)\left(2x^3y^3\right)^3$

$=\left(5x^2y^3\right)\left(2x^3y^3\right)\left(2x^3y^3\right)\left(2x^3y^3\right)$

$=(5\cdot2\cdot2\cdot2)(x\cdot x\cdot x\cdot x\cdot x\cdot x\cdot x\cdot x\cdot x\cdot x\cdot x)$

$\times(y\cdot y\cdot y\cdot y\cdot y\cdot y\cdot y\cdot y\cdot y\cdot y\cdot y\cdot y)$

$=40x^{11}y^{12}$

Chapter 2 Preview

1. The table is as follows:

$16+5=21$	$9-3=6$	$2(3)=6$	$\dfrac{24}{1}=24$
$9+2=11$	$12-8=4$	$2(26)=52$	$\dfrac{24}{24}=1$
$20+3=23$	$14-9=5$	$(13)(3)=39$	$\dfrac{120}{10}=12$
$5+9=14$	$21-10=11$	$11\cdot29=319$	$\dfrac{105}{5}=21$
$15+14=29$	$7-6=1$	$2\cdot20=40$	$\dfrac{54}{2}=27$
$30+55=85$	$18-5=13$	$59\cdot1=59$	$\dfrac{342}{3}=114$

3. $2\cdot5\cdot x=(2\cdot5)\cdot x=10x$

5. $3+2-9=5-9=5+(-9)=-4$

7. $9\cdot6+5=54+5=59$

9. $(3+5)(2+1)=(8)(3)=24$

11. $32\div4^2+75\div5^2=(32\div16)+(75\div25)$
$$=2+3$$
$$=5$$

13. $\dfrac{208}{24}$: $24\overline{)208}^{\;8\ \text{R}16}$
$$\underline{192}$$
$$16$$

15. $\dfrac{111}{8}$: $8\overline{)111}^{\;13\ \text{R}7}$
$$\underline{8\downarrow}$$
$$31$$
$$\underline{24}$$
$$7$$

17. $2\cdot2\cdot3\cdot3\cdot3=2^23^3$

19.   4 six-packs $-$ 4 cans $=$ cans left
$4\cdot6-4=24-4=20$ cans

Chapter 2 Pretest

1. $\dfrac{2 \text{ girls}}{4 \text{ children}} = \dfrac{1}{2}$

3. $\dfrac{16}{20} = \dfrac{\cancel{2} \cdot \cancel{2} \cdot 2 \cdot 2}{\cancel{2} \cdot \cancel{2} \cdot 5} = \dfrac{4}{5}$

5. $\dfrac{3}{4} \cdot \dfrac{2}{3} = \dfrac{\cancel{3} \cdot \cancel{2}}{\cancel{2} \cdot 2 \cdot \cancel{3}} = \dfrac{1}{2}$

7. $\dfrac{3}{8} \cdot 16 = \dfrac{3 \cdot 16}{8} = \dfrac{3 \cdot \cancel{8} \cdot 2}{\cancel{8}} = 6$

9. $\dfrac{32}{45} \div \dfrac{40}{63} = \dfrac{32}{45} \cdot \dfrac{63}{40} = \dfrac{\cancel{8} \cdot 4 \cdot \cancel{9} \cdot 7}{\cancel{9} \cdot 5 \cdot \cancel{8} \cdot 5} = \dfrac{28}{25}$

11. $\dfrac{5}{8} - \dfrac{1}{8} = \dfrac{5-1}{8} = \dfrac{4}{8} = \dfrac{1}{2}$

13. $\dfrac{5}{8} - \left(-\dfrac{1}{6}\right) = \dfrac{5}{8} + \dfrac{1}{6}$

$\qquad = \dfrac{5 \cdot 3}{8 \cdot 3} + \dfrac{1 \cdot 4}{6 \cdot 4}$

$\qquad = \dfrac{15}{24} + \dfrac{4}{24}$

$\qquad = \dfrac{19}{24}$

15. $\dfrac{21}{8}: \quad 8\overline{)21} \quad$ so $\quad \dfrac{21}{8} = 2 + \dfrac{5}{8} = 2\dfrac{5}{8}$

$\qquad\qquad \dfrac{16}{5}$ (with $\overset{2}{}$ quotient)

17. $12 \div 3\dfrac{1}{6} = \dfrac{12}{1} \div \dfrac{19}{6} = \dfrac{12}{1} \cdot \dfrac{6}{19} = \dfrac{72}{19} = 3\dfrac{15}{19}$

19. $\begin{array}{r} 6\dfrac{1}{6} = \\[6pt] - 3\dfrac{1}{3} = \end{array} \quad \begin{array}{r} 6\dfrac{1}{6} = \\[6pt] - 3\dfrac{1 \cdot 2}{3 \cdot 2} = \end{array} \quad \begin{array}{r} 6\dfrac{1}{6} = \\[6pt] - 3\dfrac{2}{6} = \end{array} \quad \begin{array}{r} 5\dfrac{7}{6} \\[6pt] - 3\dfrac{2}{6} \\ \hline 2\dfrac{5}{6} \end{array}$

21. $\left(\dfrac{1}{3}\right)^2 \cdot 27 + \left(\dfrac{3}{2}\right)^2 \cdot 4$

$= \left(\dfrac{1}{3}\right)\left(\dfrac{1}{3}\right) \cdot 27 + \left(\dfrac{3}{2}\right)\left(\dfrac{3}{2}\right) \cdot 4$

$= \dfrac{27}{9} + \dfrac{36}{4}$

$= 3 + 9$

$= 12$

23. $\left(1 - \dfrac{1}{4}\right)\left(1\dfrac{2}{3} - \dfrac{1}{3}\right) = \left(\dfrac{4}{4} - \dfrac{1}{4}\right)\left(\dfrac{5}{3} - \dfrac{1}{3}\right)$

$\qquad\qquad = \left(\dfrac{3}{4}\right)\left(\dfrac{4}{3}\right)$

$\qquad\qquad = 1$

25. $\dfrac{2 + \dfrac{1}{4}}{2 - \dfrac{1}{4}} = \dfrac{4\left(2 + \dfrac{1}{4}\right)}{4\left(2 - \dfrac{1}{4}\right)} = \dfrac{8 + 1}{8 - 1} = \dfrac{9}{7} = 1\dfrac{2}{7}$

27. $20 \div \dfrac{4}{5} + 1 = \dfrac{20}{1} \cdot \dfrac{5}{4} + 1$

$\qquad\qquad = 25 + 1$

$\qquad\qquad = 26$

29. $A = \dfrac{1}{2}bh = \dfrac{1}{2}(3 \text{ ft})(4 \text{ ft}) = \dfrac{12}{2} \text{ ft}^2 = 6 \text{ ft}^2$

Section 2.1

1. $\dfrac{1}{3}$: numerator = 1

3. $\dfrac{2}{3}$: numerator = 2

5. $\dfrac{x}{8}$: numerator = x

7. $\dfrac{a}{b}$: numerator = a

9. $\dfrac{2}{5}$: denominator = 5

11. $6 = \dfrac{6}{1}$: denominator = 1

13. $\dfrac{a}{12}$: denominator = 12

15. The table is as follows:

Numerator a	Denominator b	Fraction $\dfrac{a}{b}$
3	5	$\dfrac{3}{5}$
1	7	$\dfrac{1}{7}$
x	y	$\dfrac{x}{y}$
$x+1$	x	$\dfrac{x+1}{x}$

17. $\left\{ \dfrac{3}{4}, \dfrac{6}{5}, \dfrac{12}{3}, \dfrac{1}{2}, \dfrac{9}{10}, \dfrac{20}{10} \right\}$

Proper fractions: $\dfrac{3}{4}, \dfrac{1}{2}, \dfrac{9}{10}$

19. True. If a is any whole number greater than 1, a can be written as $a = \dfrac{a}{1}$, which is improper.

21. False. For example, add 1 to the numerator and denominator of the fraction $\dfrac{1}{2}$:

$$\dfrac{1+1}{2+1} = \dfrac{2}{3}, \text{ and } \dfrac{1}{2} \neq \dfrac{2}{3}.$$

23. $\dfrac{6}{8} = \dfrac{6 \div 2}{8 \div 2} = \dfrac{3}{4}$

25. $\dfrac{86}{94} = \dfrac{86 \div 2}{94 \div 2} = \dfrac{43}{47}$

27. $\dfrac{12}{9} = \dfrac{12 \div 3}{9 \div 3} = \dfrac{4}{3}$

29. $\dfrac{39}{51} = \dfrac{39 \div 3}{51 \div 3} = \dfrac{13}{17}$

31. $\dfrac{2}{3} = \dfrac{2 \cdot 2}{3 \cdot 2} = \dfrac{4}{6}$

33. $\dfrac{55}{66} = \dfrac{55 \div 11}{66 \div 11} = \dfrac{5}{6}$

35. $\dfrac{2}{3} = \dfrac{2 \cdot 4}{3 \cdot 4} = \dfrac{8}{12}$

37. $\dfrac{56}{84} = \dfrac{56 \div 7}{84 \div 7} = \dfrac{8}{12}$

39. $\dfrac{1}{6} = \dfrac{1 \cdot 2x}{6 \cdot 2x} = \dfrac{2x}{12x}$

41. $\dfrac{3}{8} = \dfrac{3 \cdot 3a}{8 \cdot 3a} = \dfrac{9a}{24a}$

43. $\dfrac{5}{6} = \dfrac{5 \cdot 4a}{6 \cdot 4a} = \dfrac{20a}{24a}$

45. Answers will vary. Because $\dfrac{1}{4} = \dfrac{1 \cdot 2}{4 \cdot 2} = \dfrac{2}{8}$,

$\dfrac{1}{4}$ is equivalent to 2 of 8 parts. Shade any 2

of the 8 parts:

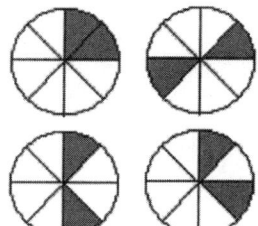

47. $\dfrac{3}{1} = 3$

49. $\dfrac{6}{3} = 2$

51. $\dfrac{37}{1} = 37$

53.

a. ▨ $\dfrac{\text{Number parts shaded}}{\text{Number equal parts}} = \dfrac{1}{2}$

b. ◥ $\dfrac{\text{Number parts shaded}}{\text{Number equal parts}} = \dfrac{1}{2}$

c. ▦ $\dfrac{\text{Number parts shaded}}{\text{Number equal parts}} = \dfrac{1}{4}$

d. ⧗ $\dfrac{\text{Number parts shaded}}{\text{Number equal parts}} = \dfrac{1}{4}$

55. Divide the distance from 0 to 1 into 4 equal parts. Then start at 0 and count over 1 part.

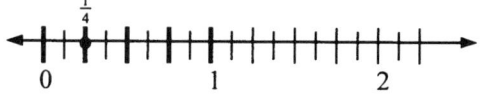

57. Divide the distance from 0 to 1 into 16 equal parts. Then start at 0 and count over 1 part.

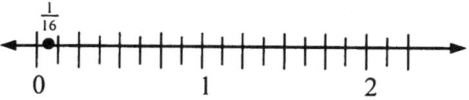

59. Divide the distance from 0 to 1 into 4 equal parts. Then start at 0 and count over 3 parts.

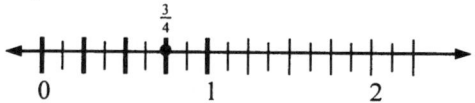

61. Divide the distance from 0 to 1 into 2 equal parts, and divide the distance from 1 to 2 into 2 equal parts. Then start at 0 and count over 3 parts.

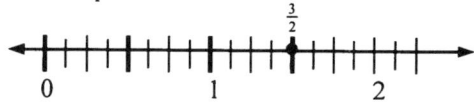

63. Divide the distance from 0 to 1 into 16 equal parts, and divide the distance from 1 to 2 into 16 equal parts. Then start at 0 and count over 31 parts.

Applying the Concepts

65. The table is as follows:

non-work-related e-mail sent	Fraction of respondents
Never	$\frac{4}{25}$
1 to 5 times a day	$\frac{47}{100}$
5 to 10 times a day	$\frac{8}{25}$
more than 10 times a day	$\frac{1}{20}$

67. $\dfrac{\text{Number of girls}}{\text{Total children}} = \dfrac{4}{5}$

69. $\dfrac{\text{Amount spent on clothes}}{\text{Total amount earned}} = \dfrac{\$2}{\$15} = \dfrac{2}{15}$

71. $\dfrac{\text{Students finished}}{\text{Students started}} = \dfrac{29}{43}$

73. $\dfrac{\text{House payment}}{\text{Total income}} = \dfrac{\$1,121}{\$1,791} = \dfrac{1,121}{1,791}$

75. Find the day on the horizontal axis, then go up to the graph and find the matching fraction value on the vertical axis.

Days since discontinuing	Fraction remaining
0	1
5	$\frac{1}{2}$
10	$\frac{1}{4}$
15	$\frac{1}{8}$
20	$\frac{1}{16}$

77. Above each day, sketch a point corresponding to the gain as measured by the vertical axis. To sketch the line graph, connect the points.

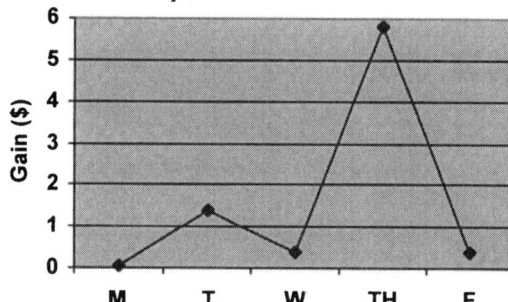

Estimating

79.

$\dfrac{1}{5}$ is closest to 0.

81.

$\dfrac{1}{8}$ is closest to 0.

Note: In comparing positive fractions, if the denominators are the same, the largest fraction has the largest numerator.

Review Problems

83. $3 + 4 \cdot 5 = 3 + 20$
$ = 23$

85. $5 \cdot 2^4 - 3 \cdot 4^2 = 5 \cdot 16 - 3 \cdot 16$
$ = 80 - 48$
$ = 32$

87. $4 \cdot 3 + 2(5 - 3) = 4 \cdot 3 + 2 \cdot 5 - 2 \cdot 3$
$ = 12 + 10 - 6$
$ = 22 - 6$
$ = 16$

89. $18 + 12 \div 4 - 3$ *divide before adding and subtracting*

$= 18 + 3 - 3$

$= 21 - 3$

$= 18$

Improving Your Quantitative Literacy

91. True. For example, $1 < 2 < 3$, and

$\dfrac{1}{4} < \dfrac{2}{4} < \dfrac{3}{4}.$

93. a.

Twins	119,648
Triplets	6,742
Quadruplets	506
+ Quintuplets and up	+ 77
	126,973

b. $\dfrac{9}{10}$ of total multiple births

$= \dfrac{9}{10} \cdot 126{,}973$

$= \dfrac{9 \cdot 126{,}973}{10}$

$= \dfrac{1{,}142{,}757}{10} \approx 114{,}276$

Yes. There were 119,648 sets of twins, which is greater than 114,276.

Section 2.2

1. 11 is prime, because the only factors of 11 are 1 and 11.

3. 105 is composite. Some factors of 105 are: 3,35,5,21,7,15.

5. 81 is composite. Some factors of 81 are: 3,9,27.

7. 13 is prime, because the only factors of 13 are 1 and 13.

9. $12 = 2 \cdot 6$
$$= 2 \cdot 2 \cdot 3 = 2^2 \cdot 3$$

11. $81 = 9 \cdot 9$
$$= 3 \cdot 3 \cdot 3 \cdot 3 = 3^4$$

13. $215 = 5 \cdot 43$

15. $15 = 3 \cdot 5$

17. $\dfrac{5}{10} = \dfrac{1 \cdot \not{5}}{2 \cdot \not{5}} = \dfrac{1}{2}$

19. $\dfrac{4}{6} = \dfrac{\not{2} \cdot 2}{\not{2} \cdot 3} = \dfrac{2}{3}$

21. $\dfrac{8}{10} = \dfrac{\not{2} \cdot 4}{\not{2} \cdot 5} = \dfrac{4}{5}$

23. $\dfrac{36}{20} = \dfrac{\not{2} \cdot \not{2} \cdot 3 \cdot 3}{\not{2} \cdot \not{2} \cdot 5} = \dfrac{9}{5}$

25. $\dfrac{42}{66} = \dfrac{\not{2} \cdot \not{3} \cdot 7}{\not{2} \cdot \not{3} \cdot 11} = \dfrac{7}{11}$

27. $\dfrac{24}{40} = \dfrac{\not{2} \cdot \not{2} \cdot \not{2} \cdot 3}{\not{2} \cdot \not{2} \cdot \not{2} \cdot 5} = \dfrac{3}{5}$

29. $\dfrac{14}{98} = \dfrac{\not{2} \cdot \not{7}}{\not{2} \cdot \not{7} \cdot 7} = \dfrac{1}{7}$

31. $\dfrac{70}{90} = \dfrac{\not{2} \cdot \not{5} \cdot 7}{\not{2} \cdot 3 \cdot 3 \cdot \not{5}} = \dfrac{7}{9}$

33. $\dfrac{42}{30} = \dfrac{\not{2} \cdot \not{3} \cdot 7}{\not{2} \cdot \not{3} \cdot 5} = \dfrac{7}{5}$

35. $\dfrac{18}{90} = \dfrac{\not{2} \cdot \not{3} \cdot \not{3}}{\not{2} \cdot \not{3} \cdot \not{3} \cdot 5} = \dfrac{1}{5}$

37. $\dfrac{110}{70} = \dfrac{\not{2} \cdot \not{5} \cdot 11}{\not{2} \cdot \not{5} \cdot 7} = \dfrac{11}{7}$

39. $\dfrac{180}{108} = \dfrac{\not{2} \cdot \not{2} \cdot \not{3} \cdot \not{3} \cdot 5}{\not{2} \cdot \not{2} \cdot \not{3} \cdot \not{3} \cdot 3} = \dfrac{5}{3}$

41. $\dfrac{96}{108} = \dfrac{\not{2} \cdot \not{2} \cdot 2 \cdot 2 \cdot 2 \cdot \not{3}}{\not{2} \cdot \not{2} \cdot \not{3} \cdot 3 \cdot 3} = \dfrac{8}{9}$

43. $\dfrac{126}{165} = \dfrac{2 \cdot \not{3} \cdot 3 \cdot 7}{\not{3} \cdot 5 \cdot 11} = \dfrac{42}{55}$

45. $\dfrac{102}{114} = \dfrac{\not{2} \cdot \not{3} \cdot 17}{\not{2} \cdot \not{3} \cdot 19} = \dfrac{17}{19}$

47. $\dfrac{294}{693} = \dfrac{2 \cdot \not{3} \cdot \not{7} \cdot 7}{\not{3} \cdot 3 \cdot \not{7} \cdot 11} = \dfrac{14}{33}$

49. a. $\dfrac{6}{51} = \dfrac{2 \cdot \cancel{3}}{\cancel{3} \cdot 17} = \dfrac{2}{17}$

b. $\dfrac{6}{52} = \dfrac{\cancel{2} \cdot 3}{\cancel{2} \cdot 2 \cdot 13} = \dfrac{3}{26}$

c. $\dfrac{6}{54} = \dfrac{\cancel{2} \cdot \cancel{3}}{\cancel{2} \cdot \cancel{3} \cdot 3 \cdot 3} = \dfrac{1}{9}$

d. $\dfrac{6}{56} = \dfrac{\cancel{2} \cdot 3}{\cancel{2} \cdot 2 \cdot 2 \cdot 7} = \dfrac{3}{28}$

e. $\dfrac{6}{57} = \dfrac{2 \cdot \cancel{3}}{\cancel{3} \cdot 19} = \dfrac{2}{19}$

51. a. $\dfrac{2}{90} = \dfrac{2 \div 2}{90 \div 2} = \dfrac{1}{45}$

b. $\dfrac{3}{90} = \dfrac{3 \div 3}{90 \div 3} = \dfrac{1}{30}$

c. $\dfrac{5}{90} = \dfrac{5 \div 5}{90 \div 5} = \dfrac{1}{18}$

d. $\dfrac{6}{90} = \dfrac{6 \div 6}{90 \div 6} = \dfrac{1}{15}$

e. $\dfrac{9}{90} = \dfrac{9 \div 9}{90 \div 9} = \dfrac{1}{10}$

53. a. $\dfrac{5}{15} = \dfrac{\cancel{5}}{3 \cdot \cancel{5}} = \dfrac{1}{3}$

Note: The slashes indicate *division* by 5.
$5 \div 5 = 1$, not 0.

b. $\dfrac{5}{6} = \dfrac{5}{2 \cdot 3} = \dfrac{5}{6}$

c. $\dfrac{6}{30} = \dfrac{\cancel{2} \cdot \cancel{3}}{\cancel{2} \cdot \cancel{3} \cdot 5} = \dfrac{1}{5}$

Note: The slashes indicate *division* by 2
and 3. $2 \div 2 = 1, 3 \div 3 = 1$, not 0.

55. $\dfrac{6}{8} = \dfrac{6 \div 2}{8 \div 2} = \dfrac{3}{4}$

$\dfrac{15}{20} = \dfrac{15 \div 5}{20 \div 5} = \dfrac{3}{4}$

$\dfrac{9}{16} = \dfrac{3 \cdot 3}{4 \cdot 4} = \dfrac{9}{16}$

$\dfrac{21}{28} = \dfrac{21 \div 7}{28 \div 7} = \dfrac{3}{4}$

$\dfrac{9}{16} \neq \dfrac{3}{4}$

57. First reduce all fractions to lowest terms:

$\dfrac{1}{2}$ is reduced $\qquad \dfrac{2}{4} = \dfrac{2 \div 2}{4 \div 2} = \dfrac{1}{2}$

$\dfrac{4}{8} = \dfrac{4 \div 4}{8 \div 4} = \dfrac{1}{2} \qquad \dfrac{8}{16} = \dfrac{8 \div 8}{16 \div 8} = \dfrac{1}{2}$

$$\tfrac{1}{2} = \tfrac{2}{4} = \tfrac{4}{8} = \tfrac{8}{16}$$

0 ———————●——————— 1

59. First reduce all fractions to lowest terms:

$\dfrac{5}{4}$ is reduced $\qquad \dfrac{10}{8} = \dfrac{10 \div 2}{8 \div 2} = \dfrac{5}{4}$

$\dfrac{20}{16} = \dfrac{20 \div 4}{16 \div 4} = \dfrac{5}{4}$

$$\tfrac{5}{4} = \tfrac{10}{8} = \tfrac{20}{16}$$

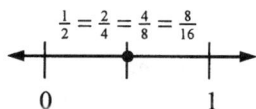

0   1   2

Applying the Concepts

61. $\dfrac{\text{Amount spent on food}}{\text{Total income}} = \dfrac{\$600}{\$2,400}$

$= \dfrac{600}{2,400}$

$= \dfrac{600 \div 600}{2,400 \div 600}$

$= \dfrac{1}{4}$

63. a. Total babies born: 4,021,726 ;

$\dfrac{1}{2}$ of total babies born $= \dfrac{1}{2} \cdot 4{,}021{,}726$

$= 2{,}010{,}863$

The number of babies born to 20-29 year olds is more than half of all babies born: $2{,}082{,}487 > 2{,}010{,}863$

b. Total babies born: 4,021,726 ;

$\dfrac{1}{3}$ of total babies born $= \dfrac{1}{3} \cdot 4{,}021{,}726$

$= \dfrac{4{,}021{,}726}{3}$

$\approx 1{,}340{,}575$

The number of babies born to 30-39 year olds is more than one-third of all babies born: $1{,}405{,}146 > 1{,}340{,}575$

65. $\dfrac{\text{Number who didn't get an A}}{\text{Total number taking the test}} = \dfrac{33-11}{33}$

$= \dfrac{22}{33}$

$= \dfrac{22 \div 11}{33 \div 11} = \dfrac{2}{3}$

67. $\dfrac{\text{Fat calories}}{\text{Total calories}} = \dfrac{70}{210} = \dfrac{70 \div 70}{210 \div 70} = \dfrac{1}{3}$

69. $\dfrac{\text{Saturated fat}}{\text{Total fat}} = \dfrac{1\text{ g}}{8\text{ g}} = \dfrac{1}{8}$

71. $\dfrac{\text{Sugar carbohydrates}}{\text{Total carbohydrates}} = \dfrac{12\text{g}}{32\text{g}} = \dfrac{12}{32}$;

$\dfrac{12}{32} = \dfrac{12 \div 4}{32 \div 4} = \dfrac{3}{8}$

Estimating

73.

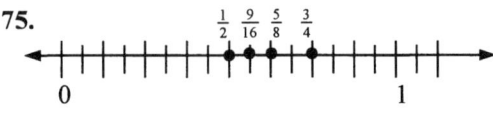

$\dfrac{5}{8}$ is closest to $\dfrac{1}{2}$.

75.

$\dfrac{9}{16}$ is closest to $\dfrac{1}{2}$.

Calculator Problems

77. Because $574 \div 7 = 82$,

$\dfrac{5}{7} = \dfrac{5 \cdot 82}{7 \cdot 82} = \dfrac{410}{574}$.

The missing term is 410.

79. Because $728 \div 8 = 91$,

$\dfrac{8}{9} = \dfrac{8 \cdot 91}{9 \cdot 91} = \dfrac{728}{819}$

The missing term is 819.

81. Because $4{,}869 \div 9 = 541$

$\dfrac{3{,}787}{4{,}869} = \dfrac{3{,}787 \div 541}{4{,}869 \div 541} = \dfrac{7}{9}$.

The missing term is 7.

83. Because $1{,}965 \div 15 = 131$

$\dfrac{262}{1{,}965} = \dfrac{262 \div 131}{1{,}965 \div 131} = \dfrac{2}{15}$.

The missing term is 2.

Review Problems

85. $6(4+8) = 6 \cdot 4 + 6 \cdot 8$

$= 24 + 48$

$= 72$

87.
$$9(3+2) = 9 \cdot 3 + 9 \cdot 2$$
$$= 27 + 18$$
$$= 45$$

Improving Your Quantitative Literacy

89. $\dfrac{\text{U.S population}}{\text{World population}} = \dfrac{293,131,160}{6,363,296,6399}$

$\dfrac{293,131,160}{6,363,296,6399}$ rounds to

$\dfrac{300,000,000}{6,000,000,000} = \dfrac{3}{60} = \dfrac{1}{20}$

Section 2.3

1. $\dfrac{2}{3}\cdot\dfrac{4}{5}=\dfrac{2\cdot 4}{3\cdot 5}=\dfrac{8}{15}$

3. $\dfrac{1}{2}\cdot\dfrac{7}{4}=\dfrac{1\cdot 7}{2\cdot 4}=\dfrac{7}{8}$

5. $\dfrac{5}{3}\cdot\dfrac{3}{5}=\dfrac{\cancel{5}\cdot\cancel{3}}{\cancel{3}\cdot\cancel{5}}=\dfrac{1}{1}=1$

7. $\dfrac{3}{4}\cdot 9=\dfrac{3}{4}\cdot\dfrac{9}{1}=\dfrac{3\cdot 9}{4\cdot 1}=\dfrac{27}{4}$

9. $\dfrac{6}{7}\left(\dfrac{7}{6}\right)=\dfrac{\cancel{6}\cdot\cancel{7}}{\cancel{7}\cdot\cancel{6}}=\dfrac{1}{1}=1$

11. $\dfrac{1}{2}\cdot\dfrac{1}{3}\cdot\dfrac{1}{4}=\dfrac{1\cdot 1\cdot 1}{2\cdot 3\cdot 4}=\dfrac{1}{24}$

13. $\dfrac{2}{5}\cdot\dfrac{3}{5}\cdot\dfrac{4}{5}=\dfrac{2\cdot 3\cdot 4}{5\cdot 5\cdot 5}=\dfrac{24}{125}$

15. $\dfrac{3}{2}\cdot\dfrac{5}{2}\cdot\dfrac{7}{2}=\dfrac{3\cdot 5\cdot 7}{2\cdot 2\cdot 2}=\dfrac{105}{8}$

17. The table is as follows:

x	y	xy
$\dfrac{1}{2}$	$\dfrac{2}{3}$	$\dfrac{1}{2}\cdot\dfrac{2}{3}=\dfrac{1\cdot 2}{2\cdot 3}=\dfrac{1}{3}$
$\dfrac{2}{3}$	$\dfrac{3}{4}$	$\dfrac{2}{3}\cdot\dfrac{3}{4}=\dfrac{2\cdot 3}{3\cdot 4}=\dfrac{2\cdot 3}{3\cdot(2\cdot 2)}=\dfrac{1}{2}$
$\dfrac{3}{4}$	$\dfrac{4}{5}$	$\dfrac{3}{4}\cdot\dfrac{4}{5}=\dfrac{3\cdot 4}{4\cdot 5}=\dfrac{3}{5}$
$\dfrac{5}{a}$	$\dfrac{a}{6}$	$\dfrac{5}{a}\cdot\dfrac{a}{6}=\dfrac{5\cdot a}{a\cdot 6}=\dfrac{5}{6}$

19. The table is as follows:

x	y	xy
$\dfrac{1}{2}$	30	$\dfrac{1}{2}\cdot\dfrac{30}{1}=\dfrac{1\cdot 30}{2\cdot 1}=\dfrac{2\cdot 3\cdot 5}{2}=\dfrac{3\cdot 5}{1}$ $=15$
$\dfrac{1}{5}$	30	$\dfrac{1}{5}\cdot\dfrac{30}{1}=\dfrac{1\cdot 30}{5\cdot 1}=\dfrac{2\cdot 3\cdot 5}{5}=\dfrac{2\cdot 3}{1}$ $=6$
$\dfrac{1}{6}$	30	$\dfrac{1}{6}\cdot\dfrac{30}{1}=\dfrac{1\cdot 30}{6\cdot 1}=\dfrac{2\cdot 3\cdot 5}{2\cdot 3}=\dfrac{5}{1}$ $=5$
$\dfrac{1}{15}$	30	$\dfrac{1}{15}\cdot\dfrac{30}{1}=\dfrac{1\cdot 30}{15\cdot 1}=\dfrac{2\cdot 3\cdot 5}{3\cdot 5}=\dfrac{2}{1}$ $=2$

21. $\dfrac{9}{20}\cdot\dfrac{4}{3}=\dfrac{9\cdot 4}{20\cdot 3}$

$\quad=\dfrac{(\cancel{3}\cdot 3)(\cancel{2}\cdot\cancel{2})}{(\cancel{2}\cdot\cancel{2}\cdot 5)(\cancel{3})}$

$\quad=\dfrac{3}{5}$

23. $\dfrac{3}{4}\cdot 12=\dfrac{3}{4}\cdot\dfrac{12}{1}$

$\quad=\dfrac{3\cdot 12}{4\cdot 1}$

$\quad=\dfrac{3(\cancel{2}\cdot\cancel{2}\cdot 3)}{\cancel{2}\cdot\cancel{2}}$

$\quad=9$

25. $\dfrac{1}{3}(3)=\dfrac{1}{3}\cdot\dfrac{3}{1}$

$\quad=\dfrac{1\cdot\cancel{3}}{\cancel{3}\cdot 1}$

$\quad=1$

27. $\dfrac{2}{5} \cdot 20 = \dfrac{2}{5} \cdot \dfrac{20}{1}$

$= \dfrac{2 \cdot (2 \cdot 2 \cdot \cancel{5})}{\cancel{5} \cdot 1}$

$= \dfrac{8}{1}$

$= 8$

29. $\dfrac{72}{35} \cdot \dfrac{55}{108} \cdot \dfrac{7}{110}$

$= \dfrac{(\cancel{2} \cdot \cancel{2} \cdot \cancel{2} \cdot \cancel{3} \cdot \cancel{3}) \cdot (\cancel{5} \cdot \cancel{11}) \cdot \cancel{7}}{(\cancel{5} \cdot \cancel{7})(\cancel{2} \cdot \cancel{2} \cdot \cancel{3} \cdot \cancel{3} \cdot 3)(\cancel{2} \cdot 5 \cdot \cancel{11})}$

$= \dfrac{1}{15}$

31. $\left(\dfrac{2}{3}\right)^2 = \dfrac{2}{3} \cdot \dfrac{2}{3}$

$= \dfrac{2 \cdot 2}{3 \cdot 3}$

$= \dfrac{4}{9}$

33. $\left(\dfrac{3}{4}\right)^2 = \dfrac{3}{4} \cdot \dfrac{3}{4}$

$= \dfrac{3 \cdot 3}{4 \cdot 4}$

$= \dfrac{9}{16}$

35. $\left(\dfrac{1}{2}\right)^2 = \left(\dfrac{1}{2}\right)\left(\dfrac{1}{2}\right)$

$= \dfrac{1 \cdot 1}{2 \cdot 2}$

$= \dfrac{1}{4}$

37. $\left(\dfrac{2}{3}\right)^3 = \left(\dfrac{2}{3}\right)\left(\dfrac{2}{3}\right)\left(\dfrac{2}{3}\right)$

$= \dfrac{2 \cdot 2 \cdot 2}{3 \cdot 3 \cdot 3}$

$= \dfrac{8}{27}$

39. $\left(\dfrac{3}{4}\right)^2 \cdot \left(\dfrac{8}{9}\right) = \dfrac{3}{4} \cdot \dfrac{3}{4} \cdot \dfrac{8}{9}$

$= \dfrac{3 \cdot 3 \cdot 8}{4 \cdot 4 \cdot 9}$

$= \dfrac{\cancel{3} \cdot \cancel{3} \cdot \cancel{2} \cdot \cancel{2} \cdot \cancel{2}}{(\cancel{2} \cdot \cancel{2}) \cdot (\cancel{2} \cdot 2) \cdot (\cancel{3} \cdot \cancel{3})}$

$= \dfrac{1}{2}$

41. $\left(\dfrac{1}{2}\right)^2 \left(\dfrac{3}{5}\right)^2 = \dfrac{1}{2} \cdot \dfrac{1}{2} \cdot \dfrac{3}{5} \cdot \dfrac{3}{5}$

$= \dfrac{1 \cdot 1 \cdot 3 \cdot 3}{2 \cdot 2 \cdot 5 \cdot 5}$

$= \dfrac{9}{100}$

43. $\left(\dfrac{1}{2}\right)^2 \cdot 8 + \left(\dfrac{1}{3}\right)^2 \cdot 9$

$= \dfrac{1}{2} \cdot \dfrac{1}{2} \cdot \dfrac{8}{1} + \dfrac{1}{3} \cdot \dfrac{1}{3} \cdot \dfrac{9}{1}$

$= \dfrac{\cancel{2} \cdot \cancel{2} \cdot 2}{\cancel{2} \cdot \cancel{2}} + \dfrac{\cancel{3} \cdot \cancel{3}}{\cancel{3} \cdot \cancel{3}}$

$= 2 + 1$

$= 3$

45. $\dfrac{3}{8}$ of $64 = \dfrac{3}{8} \cdot 64$

$\quad = \dfrac{3}{8} \cdot \dfrac{64}{1}$

$\quad = \dfrac{3 \cdot 64}{8 \cdot 1}$

$\quad = \dfrac{3 \cdot \left(\cancel{2} \cdot \cancel{2} \cdot \cancel{2} \cdot 2 \cdot 2 \cdot 2\right)}{\left(\cancel{2} \cdot \cancel{2} \cdot \cancel{2}\right) \cdot 1}$

$\quad = \dfrac{24}{1}$

$\quad = 24$

47. $\dfrac{1}{3}(8+4) = \dfrac{1}{3}(12)$

$\quad = \dfrac{1}{3} \cdot \dfrac{12}{1}$

$\quad = \dfrac{1 \cdot 12}{3 \cdot 1}$

$\quad = \dfrac{1 \cdot \left(2 \cdot 2 \cdot \cancel{3}\right)}{\cancel{3} \cdot 1}$

$\quad = \dfrac{4}{1}$

$\quad = 4$

49. $\dfrac{1}{2}$ of $\dfrac{3}{4}$ of $24 = \dfrac{1}{2} \cdot \dfrac{3}{4} \cdot \dfrac{24}{1}$

$\quad = \dfrac{1 \cdot 3 \cdot 24}{2 \cdot 4 \cdot 1}$

$\quad = \dfrac{1 \cdot 3 \cdot \left(\cancel{2} \cdot \cancel{2} \cdot \cancel{2} \cdot 3\right)}{\cancel{2} \cdot \left(\cancel{2} \cdot \cancel{2}\right) \cdot 1}$

$\quad = \dfrac{9}{1}$

$\quad = 9$

51. $\dfrac{1}{2} \cdot \dfrac{3}{5} = \dfrac{1 \cdot 3}{2 \cdot 5} = \dfrac{3}{10}$

Note: The mistake in the original problem was that the numerators were added, not multiplied.

53. a. The table is as follows:

x	x^2
1	$1^2 = 1 \cdot 1 = 1$
2	$2^2 = 2 \cdot 2 = 4$
3	$3^2 = 3 \cdot 3 = 9$
4	$4^2 = 4 \cdot 4 = 16$
5	$5^2 = 5 \cdot 5 = 25$
6	$6^2 = 6 \cdot 6 = 36$
7	$7^2 = 7 \cdot 7 = 49$
8	$8^2 = 8 \cdot 8 = 64$

b. For numbers larger than 1, the square of a number is *larger* than the number.

55. $A = \dfrac{1}{2}bh$

$\quad = \dfrac{1}{2} \cdot 19 \text{ in} \cdot 14 \text{ in}$

$\quad = \dfrac{1 \cdot 19 \cdot 14}{2} \text{ in}^2$

$\quad = \dfrac{1 \cdot 19 \cdot \cancel{2} \cdot 7}{\cancel{2}} \text{ in}^2$

$\quad = 133 \text{ in}^2$

57. $A = \dfrac{1}{2}bh$

$\quad = \dfrac{1}{2} \cdot \dfrac{4}{3} \text{ ft} \cdot \dfrac{2}{3} \text{ ft}$

$\quad = \dfrac{1 \cdot 4 \cdot \cancel{2}}{\cancel{2} \cdot 3 \cdot 3} \text{ ft}^2$

$\quad = \dfrac{4}{9} \text{ ft}^2$

59. $A = \dfrac{1}{2}bh$

$\quad = \dfrac{1}{2} \cdot 2\text{yd} \cdot 3\text{yd}$

$\quad = \dfrac{1 \cdot \cancel{2} \cdot 3}{\cancel{2}} \text{ yd}^2$

$\quad = 3 \text{ yd}^2$

61.

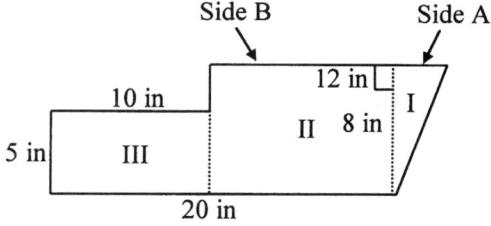

Separate the figure into 3 parts.
Find the length of the missing sides:

Side A = 10 in + 12 in − 20 in = 2 in.

Side B = 12 in − 2 in = 10 in

$$\text{Area I} = \frac{1}{2}bh \qquad \text{Area II} = lw$$

$$= \frac{1}{2} \cdot 2 \text{ in} \cdot 8 \text{ in} \qquad\qquad = 10 \text{ in} \cdot 8 \text{ in}$$

$$\qquad\qquad\qquad\qquad = 80 \text{ in}^2$$

$$= 1 \text{ in} \cdot 8 \text{ in}$$

$$= 8 \text{ in}^2$$

$$\text{Area III} = lw$$

$$= 10 \text{ in} \cdot 5 \text{ in}$$

$$= 50 \text{ in}^2$$

$$\text{Area} = \text{Area I} + \text{Area II} + \text{Area III}$$

$$= 8 \text{ in}^2 + 80 \text{ in}^2 + 50 \text{ in}^2$$

$$= 138 \text{ in}^2$$

Applying the Concepts

63. $\dfrac{11}{50}$ of $15,800 = \dfrac{11}{50} \cdot 15,800$

$$= \frac{11}{50} \cdot \frac{15,800}{1}$$

$$= \frac{11 \cdot \left(\cancel{2} \cdot 2 \cdot 2 \cdot \cancel{5} \cdot \cancel{5} \cdot 79\right)}{\cancel{2} \cdot \cancel{5} \cdot \cancel{5}}$$

$$= 3,476$$

3,476 students.

65. $\dfrac{6}{5}$ of volume of Rally 105

$$= \frac{6}{5} \text{ of } 105,400$$

$$= \frac{6}{5} \cdot \frac{105,400}{1}$$

$$= \frac{6 \cdot \cancel{5} \cdot 21,080}{\cancel{5} \cdot 1}$$

$$= 126,480 \approx 126,500 \text{ (to the nearest hundred)}$$

The volume of the Rally 126 is 126,500 ft³.

67. First, find the number of children who were wearing helmets:

$$\frac{2}{5} \text{ of children}$$

$$= \frac{2}{5} \text{ of } 8,159$$

$$= \frac{2 \cdot 8159}{5}$$

$$= \frac{16,318}{5} \approx 3,264 \text{ wore helmets}$$

For those who wore helmets, find the number wearing them correctly:

$$\frac{13}{20} \text{ of helmeted}$$

$$= \frac{13}{20} \cdot 3,264$$

$$= \frac{13 \cdot \cancel{4} \cdot 816}{\cancel{4} \cdot 5}$$

$$= \frac{10,608}{5} \approx 2,122$$

2,122 children wore helmets correctly.

Estimating

69. $\dfrac{11}{5} \cdot \dfrac{19}{20}$ rounds to $\dfrac{10}{5} \cdot \dfrac{20}{20} = 2 \cdot 1 = 2$

71. $\dfrac{16}{5} \cdot \dfrac{23}{24}$ rounds to $\dfrac{15}{5} \cdot \dfrac{24}{24} = 3 \cdot 1 = 3$

73. $\dfrac{1}{8} \cdot \dfrac{15}{32}$ rounds to $\dfrac{0}{8} \cdot \dfrac{16}{32} = 0 \cdot \dfrac{1}{2} = 0$

Review Problems

75. $6 \div 2 = 3$

77. $8 \div 4 = 2$

79. $15 \div 3 = 5$

81. $18 \div 6 = 3$

Extending the Concepts

83. $1, \dfrac{1}{3}, \dfrac{1}{9}, \ldots$

The first number in the sequence is 1. Note that $1 \cdot \dfrac{1}{3} = \dfrac{1}{3}$ and $\dfrac{1}{3} \cdot \dfrac{1}{3} = \dfrac{1}{9}$, i.e. each number after 1 comes from multiplying the previous number by $\dfrac{1}{3}$. The next number in the sequence is $\dfrac{1}{9} \cdot \dfrac{1}{3} = \dfrac{1}{27}$

85. $\dfrac{3}{2}, 1, \dfrac{2}{3}, \dfrac{4}{9}, \ldots$

The first number in the sequence is $\dfrac{3}{2}$.

Note that $\dfrac{3}{2} \cdot \dfrac{2}{3} = 1$, $1 \cdot \dfrac{2}{3} = \dfrac{2}{3}$ and $\dfrac{2}{3} \cdot \dfrac{2}{3} = \dfrac{4}{9}$,

i.e. each number after $\dfrac{3}{2}$ comes from multiplying the previous number by $\dfrac{2}{3}$.

The next number in the sequence is $\dfrac{4}{9} \cdot \dfrac{2}{3} = \dfrac{8}{27}$

Improving your Quantitative Literacy

87. $\dfrac{3}{4}$ of 55 companies

$= \dfrac{3}{4} \cdot \dfrac{55}{1}$

$= \dfrac{165}{4}$

$= 41\ \text{R}1$

Since ¾ of 55 is more than 41, at least 42 of the companies will cut spending.

Section 2.4

1. $\dfrac{3}{4} \div \dfrac{1}{5} = \dfrac{3}{4} \cdot \dfrac{5}{1}$

$\qquad = \dfrac{3 \cdot 5}{4 \cdot 1}$

$\qquad = \dfrac{15}{4}$

3. $\dfrac{2}{3} \div \dfrac{1}{2} = \dfrac{2}{3} \cdot \dfrac{2}{1}$

$\qquad = \dfrac{4}{3}$

5. $6 \div \dfrac{2}{3} = 6 \cdot \dfrac{3}{2}$

$\qquad = \dfrac{6 \cdot 3}{2}$

$\qquad = \dfrac{3 \cdot \cancel{2} \cdot 3}{\cancel{2}}$

$\qquad = 9$

7. $20 \div \dfrac{1}{10} = 20 \cdot \dfrac{10}{1}$

$\qquad = \dfrac{20 \cdot 10}{1}$

$\qquad = 200$

9. $\dfrac{3}{4} \div 2 = \dfrac{3}{4} \cdot \dfrac{1}{2}$

$\qquad = \dfrac{3 \cdot 1}{4 \cdot 2}$

$\qquad = \dfrac{3}{8}$

11. $\dfrac{7}{8} \div \dfrac{7}{8} = \dfrac{7}{8} \cdot \dfrac{8}{7}$

$\qquad = \dfrac{\cancel{7} \cdot \cancel{8}}{\cancel{8} \cdot \cancel{7}}$

$\qquad = 1$

13. $\dfrac{7}{8} \div \dfrac{8}{7} = \dfrac{7}{8} \cdot \dfrac{7}{8}$

$\qquad = \dfrac{7 \cdot 7}{8 \cdot 8}$

$\qquad = \dfrac{49}{64}$

15. $\dfrac{9}{16} \div \dfrac{3}{4} = \dfrac{9}{16} \cdot \dfrac{4}{3}$

$\qquad = \dfrac{9 \cdot 4}{16 \cdot 3}$

$\qquad = \dfrac{\cancel{3} \cdot 3 \cdot \cancel{4}}{\cancel{4} \cdot 4 \cdot \cancel{3}}$

$\qquad = \dfrac{3}{4}$

17. $\dfrac{25}{46} \div \dfrac{40}{69} = \dfrac{25}{46} \cdot \dfrac{69}{40}$

$\qquad = \dfrac{\cancel{5} \cdot 5 \cdot \cancel{23} \cdot 3}{\cancel{23} \cdot 2 \cdot \cancel{5} \cdot 8}$

$\qquad = \dfrac{15}{16}$

19. $\dfrac{13}{28} \div \dfrac{39}{14} = \dfrac{13}{28} \cdot \dfrac{14}{39}$

$\qquad = \dfrac{\cancel{13} \cdot \cancel{14}}{\cancel{14} \cdot 2 \cdot \cancel{13} \cdot 3}$

$\qquad = \dfrac{1}{6}$

21. $\dfrac{27}{196} \div \dfrac{9}{392} = \dfrac{27}{196} \cdot \dfrac{392}{9}$

$\qquad = \dfrac{3 \cdot \cancel{9} \cdot \cancel{196} \cdot 2}{\cancel{196} \cdot \cancel{9}}$

$\qquad = 6$

23. $\dfrac{25}{18} \div 5 = \dfrac{25}{18} \div \dfrac{5}{1}$

$\qquad = \dfrac{25}{18} \cdot \dfrac{1}{5}$

$\qquad = \dfrac{5 \cdot \cancel{5}}{18 \cdot \cancel{5}}$

$\qquad = \dfrac{5}{18}$

25. $6 \div \dfrac{4}{3} = \dfrac{6}{1} \div \dfrac{4}{3}$

$\qquad = \dfrac{6}{1} \cdot \dfrac{3}{4}$

$\qquad = \dfrac{\cancel{2} \cdot 3 \cdot 3}{1 \cdot 2 \cdot \cancel{2}}$

$\qquad = \dfrac{9}{2}$

27. $\dfrac{4}{3} \div 6 = \dfrac{4}{3} \div \dfrac{6}{1}$

$\qquad = \dfrac{4}{3} \cdot \dfrac{1}{6}$

$\qquad = \dfrac{2 \cdot \cancel{2}}{3 \cdot \cancel{2} \cdot 3}$

$\qquad = \dfrac{2}{9}$

29. $10 \div \left(\dfrac{1}{2}\right)^2 = 10 \div \dfrac{1}{4}$

$\qquad = 10 \cdot \dfrac{4}{1}$

$\qquad = 40$

31. $\dfrac{18}{35} \div \left(\dfrac{6}{7}\right)^2 = \dfrac{18}{35} \div \dfrac{36}{49}$

$\qquad = \dfrac{18}{35} \cdot \dfrac{49}{36}$

$\qquad = \dfrac{\cancel{18} \cdot \cancel{7} \cdot 7}{\cancel{7} \cdot 5 \cdot \cancel{18} \cdot 2}$

$\qquad = \dfrac{7}{10}$

33. $\dfrac{4}{5} \div \dfrac{1}{10} + 5 = \dfrac{4}{5} \cdot \dfrac{10}{1} + 5$

$\qquad = 8 + 5$

$\qquad = 13$

35. $10 + \dfrac{11}{12} \div \dfrac{11}{24} = 10 + \dfrac{11}{12} \cdot \dfrac{24}{11}$

$\qquad = 10 + 2$

$\qquad = 12$

37. $24 \div \left(\dfrac{2}{5}\right)^2 + 25 \div \left(\dfrac{5}{6}\right)^2 = 24 \div \dfrac{4}{25} + 25 \div \dfrac{25}{36}$

$\qquad = 24 \cdot \dfrac{25}{4} + 25 \cdot \dfrac{36}{25}$

$\qquad = 6 \cdot 25 + 1 \cdot 36$

$\qquad = 150 + 36$

$\qquad = 186$

39. $100 \div \left(\dfrac{5}{7}\right)^2 + 200 \div \left(\dfrac{2}{3}\right)^2$

$\qquad = 100 \div \dfrac{25}{49} + 200 \div \dfrac{4}{9}$

$\qquad = 100 \cdot \dfrac{49}{25} + 200 \cdot \dfrac{9}{4}$

$\qquad = 4 \cdot 49 + 50 \cdot 9$

$\qquad = 196 + 450$

$\qquad = 646$

41. $\dfrac{3}{8} \div \dfrac{5}{8} = \dfrac{3}{8} \cdot \dfrac{8}{5}$

$\qquad = \dfrac{3}{5}$

43. $\left(18 \div \dfrac{3}{5}\right) + 10 = 18 \cdot \dfrac{5}{3} + 10$

$\qquad\qquad = 6 \cdot 5 + 10$

$\qquad\qquad = 30 + 10$

$\qquad\qquad = 40$

The number is 40.

45. $3 \div \dfrac{1}{5} = 3 \cdot \dfrac{5}{1} = 3 \cdot 5$

Applying the Concepts

47. Total yds ÷ yds per blanket = blankets

$12 \div \dfrac{6}{7} = 12 \cdot \dfrac{7}{6}$

$\qquad = 2 \cdot 7$

$\qquad = 14$

14 blankets can be made.

49. Total lbs ÷ lbs per bag = bags

$12 \div \dfrac{1}{4} = 12 \cdot \dfrac{4}{1}$

$\qquad = 48$

48 bags can be filled.

51. Total tsp ÷ tsp per spoon = spoons

$\dfrac{3}{4} \div \dfrac{1}{8} = \dfrac{3}{4} \cdot \dfrac{8}{1}$

$\qquad = 3 \cdot 2$

$\qquad = 6$

He will need to fill 6 spoons.

53. $\dfrac{\text{Number of females}}{\text{Number of students}} = \dfrac{14}{32}$

$\dfrac{14}{32} = \dfrac{2 \cdot 7}{2 \cdot 16} = \dfrac{7}{16}$

55. From problem #53, $\frac{7}{16}$ of the students are female.

$\dfrac{7}{16}$ of total students $= \dfrac{7}{16} \cdot 4{,}064$

$\qquad\qquad = \dfrac{7 \cdot \cancel{16} \cdot 254}{\cancel{16}}$

$\qquad\qquad = 1{,}778$

1,778 of the students are female.

57. Total pints ÷ pints per carton = cartons

$14 \div \dfrac{1}{2} = 14 \cdot \dfrac{2}{1}$

$\qquad = 28$

28 cartons can be filled.

Estimating

59. $\dfrac{11}{5} \div \dfrac{19}{20}$ rounds to $\dfrac{10}{5} \div \dfrac{20}{20} = 2 \div 1 = 2$

61. $\dfrac{15}{8} \div \dfrac{23}{24}$ rounds to $\dfrac{16}{8} \div \dfrac{24}{24} = 2 \div 1 = 2$

63. $\dfrac{1}{2} \div 40 = \dfrac{1}{2} \cdot \dfrac{1}{40}$ rounds to $\dfrac{1}{2} \cdot 0 = 0$

Review Problems

65. $\dfrac{1}{2} = \dfrac{1 \cdot 3}{2 \cdot 3} = \dfrac{3}{6}$

67. $\dfrac{3}{2} = \dfrac{3 \cdot 3}{2 \cdot 3} = \dfrac{9}{6}$

69. $\dfrac{1}{3} = \dfrac{1 \cdot 4}{3 \cdot 4} = \dfrac{4}{12}$

71. $\dfrac{2}{3} = \dfrac{2 \cdot 4}{3 \cdot 4} = \dfrac{8}{12}$

Section 2.5

1. $\dfrac{3}{6}+\dfrac{1}{6}=\dfrac{3+1}{6}$

$=\dfrac{4}{6}$

$=\dfrac{2}{3}$

3. $\dfrac{5}{8}-\dfrac{3}{8}=\dfrac{5-3}{8}$

$=\dfrac{2}{8}$

$=\dfrac{1}{4}$

5. $\dfrac{3}{4}-\dfrac{1}{4}=\dfrac{3-1}{4}$

$=\dfrac{2}{4}$

$=\dfrac{1}{2}$

7. $\dfrac{2}{3}-\dfrac{1}{3}=\dfrac{2-1}{3}=\dfrac{1}{3}$

9. $\dfrac{1}{4}+\dfrac{2}{4}+\dfrac{3}{4}=\dfrac{1+2+3}{4}$

$=\dfrac{6}{4}$

$=\dfrac{3}{2}$

11. $\dfrac{x+7}{2}-\dfrac{1}{2}=\dfrac{x+7-1}{2}$

$=\dfrac{x+6}{2}$

13. $\dfrac{1}{10}+\dfrac{3}{10}+\dfrac{4}{10}=\dfrac{1+3+4}{10}$

$=\dfrac{8}{10}$

$=\dfrac{4}{5}$

15. $\dfrac{1}{3}+\dfrac{4}{3}+\dfrac{5}{3}=\dfrac{1+4+5}{3}$

$=\dfrac{10}{3}$

17. The table is as follows:

a	b	$a+b$
$\dfrac{1}{2}$	$\dfrac{1}{3}$	$\dfrac{1}{2}+\dfrac{1}{3}=\dfrac{1\cdot 3}{2\cdot 3}+\dfrac{1\cdot 2}{3\cdot 2}$ $=\dfrac{3}{6}+\dfrac{2}{6}=\dfrac{5}{6}$
$\dfrac{1}{3}$	$\dfrac{1}{4}$	$\dfrac{1}{3}+\dfrac{1}{4}=\dfrac{1\cdot 4}{3\cdot 4}+\dfrac{1\cdot 3}{4\cdot 3}$ $=\dfrac{4}{12}+\dfrac{3}{12}=\dfrac{7}{12}$
$\dfrac{1}{4}$	$\dfrac{1}{5}$	$\dfrac{1}{4}+\dfrac{1}{5}=\dfrac{1\cdot 5}{4\cdot 5}+\dfrac{1\cdot 4}{5\cdot 4}$ $=\dfrac{5}{20}+\dfrac{4}{20}=\dfrac{9}{20}$
$\dfrac{1}{5}$	$\dfrac{1}{6}$	$\dfrac{1}{5}+\dfrac{1}{6}=\dfrac{1\cdot 6}{5\cdot 6}+\dfrac{1\cdot 5}{6\cdot 5}$ $=\dfrac{6}{30}+\dfrac{5}{30}=\dfrac{11}{30}$

19. The table is as follows:

a	b	$a+b$
$\dfrac{1}{12}$	$\dfrac{1}{2}$	$\dfrac{1}{12}+\dfrac{1}{2}=\dfrac{1}{12}+\dfrac{1\cdot 6}{2\cdot 6}$ $=\dfrac{1}{12}+\dfrac{6}{12}=\dfrac{7}{12}$
$\dfrac{1}{12}$	$\dfrac{1}{3}$	$\dfrac{1}{12}+\dfrac{1}{3}=\dfrac{1}{12}+\dfrac{1\cdot 4}{3\cdot 4}$ $=\dfrac{1}{12}+\dfrac{4}{12}=\dfrac{5}{12}$
$\dfrac{1}{12}$	$\dfrac{1}{4}$	$\dfrac{1}{12}+\dfrac{1}{4}=\dfrac{1}{12}+\dfrac{1\cdot 3}{4\cdot 3}$ $=\dfrac{1}{12}+\dfrac{3}{12}=\dfrac{4}{12}=\dfrac{1}{3}$
$\dfrac{1}{12}$	$\dfrac{1}{6}$	$\dfrac{1}{12}+\dfrac{1}{6}=\dfrac{1}{12}+\dfrac{1\cdot 2}{6\cdot 2}$ $=\dfrac{1}{12}+\dfrac{2}{12}=\dfrac{3}{12}=\dfrac{1}{4}$

21.
$$\left.\begin{array}{l} 9=3\cdot 3 \\ 3=3 \end{array}\right\} \text{LCD}=3\cdot 3=9$$
$$\frac{4}{9}+\frac{1}{3}=\frac{4}{9}+\frac{1\cdot 3}{3\cdot 3}$$
$$=\frac{4}{9}+\frac{3}{9}$$
$$=\frac{7}{9}$$

23. $\text{LCD}=1\cdot 3=3$
$$2+\frac{1}{3}=\frac{2\cdot 3}{1\cdot 3}+\frac{1}{3}$$
$$=\frac{6}{3}+\frac{1}{3}$$
$$=\frac{7}{3}$$

25. $\text{LCD}=4\cdot 1=4$
$$\frac{3}{4}+1=\frac{3}{4}+\frac{1\cdot 4}{1\cdot 4}$$
$$=\frac{3}{4}+\frac{4}{4}$$
$$=\frac{7}{4}$$

27. $\text{LCD}=2\cdot 3=6$
$$\frac{1}{2}+\frac{2}{3}=\frac{1\cdot 3}{2\cdot 3}+\frac{2\cdot 2}{3\cdot 2}$$
$$=\frac{3}{6}+\frac{4}{6}$$
$$=\frac{7}{6}$$

29. $\text{LCD}=4\cdot 5=20$
$$\frac{1}{4}+\frac{1}{5}=\frac{1\cdot 5}{4\cdot 5}+\frac{1\cdot 4}{5\cdot 4}$$
$$=\frac{5}{20}+\frac{4}{20}$$
$$=\frac{9}{20}$$

31. $\text{LCD}=2\cdot 5=10$
$$\frac{1}{2}+\frac{1}{5}=\frac{1\cdot 5}{2\cdot 5}+\frac{1\cdot 2}{5\cdot 2}$$
$$=\frac{5}{10}+\frac{2}{10}$$
$$=\frac{7}{10}$$

33.
$$\left.\begin{array}{l} 12=2\cdot 2\cdot 3 \\ 8=2\cdot 2\cdot 2 \end{array}\right\} \text{LCD}=2\cdot 2\cdot 2\cdot 3=24$$
$$\frac{5}{12}+\frac{3}{8}=\frac{5\cdot 2}{12\cdot 2}+\frac{3\cdot 3}{8\cdot 3}$$
$$=\frac{10}{24}+\frac{9}{24}$$
$$=\frac{19}{24}$$

35.
$$\left.\begin{array}{l} 20=2\cdot 2\cdot 5 \\ 30=2\cdot 3\cdot 5 \end{array}\right\} \text{LCD}=2\cdot 2\cdot 3\cdot 5=60$$
$$\frac{8}{30}-\frac{1}{20}=\frac{8\cdot 2}{30\cdot 2}-\frac{1\cdot 3}{20\cdot 3}$$
$$=\frac{16}{60}-\frac{3}{60}$$
$$=\frac{13}{60}$$

37.
$$\left.\begin{array}{l}10 = 2\cdot 5\\100 = 2\cdot 2\cdot 5\cdot 5\end{array}\right\}\text{LCD} = 2\cdot 2\cdot 5\cdot 5 = 100$$

$$\begin{aligned}\frac{3}{10}+\frac{1}{100}&=\frac{3\cdot 10}{10\cdot 10}+\frac{1}{100}\\&=\frac{30}{100}+\frac{1}{100}\\&=\frac{31}{100}\end{aligned}$$

39.
$$\left.\begin{array}{l}36 = 2\cdot 2\cdot 3\cdot 3\\48 = 2\cdot 2\cdot 2\cdot 2\cdot 3\end{array}\right\}\text{LCD} = 2\cdot 2\cdot 2\cdot 2\cdot 3\cdot 3 = 144$$

$$\begin{aligned}\frac{10}{36}+\frac{9}{48}&=\frac{10\cdot 4}{36\cdot 4}+\frac{9\cdot 3}{48\cdot 3}\\&=\frac{40}{144}+\frac{27}{144}\\&=\frac{67}{144}\end{aligned}$$

41.
$$\left.\begin{array}{l}30 = 2\cdot 3\cdot 5\\42 = 2\cdot 3\cdot 7\end{array}\right\}\text{LCD} = 2\cdot 3\cdot 5\cdot 7 = 210$$

$$\begin{aligned}\frac{17}{30}+\frac{11}{42}&=\frac{17\cdot 7}{30\cdot 7}+\frac{11\cdot 5}{42\cdot 5}\\&=\frac{119}{210}+\frac{55}{210}\\&=\frac{174}{210}\\&=\frac{29}{35}\end{aligned}$$

43.
$$\left.\begin{array}{l}84 = 2\cdot 2\cdot 3\cdot 7\\90 = 2\cdot 3\cdot 3\cdot 5\end{array}\right\}\text{LCD} = 2\cdot 2\cdot 3\cdot 3\cdot 5\cdot 7 = 1{,}260$$

$$\begin{aligned}\frac{25}{84}+\frac{41}{90}&=\frac{25\cdot 15}{84\cdot 15}+\frac{41\cdot 14}{90\cdot 14}\\&=\frac{375}{1{,}260}+\frac{574}{1{,}260}\\&=\frac{949}{1{,}260}\end{aligned}$$

45.
$$\left.\begin{array}{l}126 = 2\cdot 3\cdot 3\cdot 7\\180 = 2\cdot 2\cdot 3\cdot 3\cdot 5\end{array}\right\}\text{LCD} = 2\cdot 3\cdot 3\cdot 3\cdot 5\cdot 7 = 1{,}260$$

$$\begin{aligned}\frac{13}{126}-\frac{13}{180}&=\frac{13\cdot 10}{126\cdot 10}-\frac{13\cdot 7}{180\cdot 7}\\&=\frac{130}{1{,}260}-\frac{91}{1{,}260}\\&=\frac{39}{1{,}260}\\&=\frac{13}{420}\end{aligned}$$

47.
$$\left.\begin{array}{l}4 = 2\cdot 2\\8 = 2\cdot 2\cdot 2\\6 = 2\cdot 3\end{array}\right\}\text{LCD} = 2\cdot 2\cdot 2\cdot 3 = 24$$

$$\begin{aligned}\frac{3}{4}+\frac{1}{8}+\frac{5}{6}&=\frac{3\cdot 6}{4\cdot 6}+\frac{1\cdot 3}{8\cdot 3}+\frac{5\cdot 4}{6\cdot 4}\\&=\frac{18}{24}+\frac{3}{24}+\frac{20}{24}\\&=\frac{41}{24}\end{aligned}$$

49.
$$\left.\begin{array}{l}10 = 2\cdot 5\\12 = 2\cdot 2\cdot 3\\6 = 2\cdot 3\end{array}\right\}\text{LCD} = 2\cdot 2\cdot 3\cdot 5 = 60$$

$$\begin{aligned}\frac{3}{10}+\frac{5}{12}+\frac{1}{6}&=\frac{3\cdot 6}{10\cdot 6}+\frac{5\cdot 5}{12\cdot 5}+\frac{1\cdot 10}{6\cdot 10}\\&=\frac{18}{60}+\frac{25}{60}+\frac{10}{60}\\&=\frac{53}{60}\end{aligned}$$

51.

$$\left.\begin{array}{l} 2 = 2 \\ 3 = 3 \\ 4 = 2\cdot 2 \\ 6 = 2\cdot 3 \end{array}\right\} LCD = 2\cdot 2\cdot 3 = 12$$

$$\frac{1}{2}+\frac{1}{3}+\frac{1}{4}+\frac{1}{6}=\frac{1\cdot 6}{2\cdot 6}+\frac{1\cdot 4}{3\cdot 4}+\frac{1\cdot 3}{4\cdot 3}+\frac{1\cdot 2}{6\cdot 2}$$

$$=\frac{6}{12}+\frac{4}{12}+\frac{3}{12}+\frac{2}{12}$$

$$=\frac{15}{12}$$

$$=\frac{5}{4}$$

53.

$$\left.\begin{array}{l} 7 = 7 \\ 9 = 3\cdot 3 \end{array}\right\} LCD = 7\cdot 3\cdot 3 = 63$$

$$\frac{3}{7}+2+\frac{1}{9}=\frac{3\cdot 9}{7\cdot 9}+\frac{2\cdot 63}{1\cdot 63}+\frac{1\cdot 7}{9\cdot 7}$$

$$=\frac{27}{63}+\frac{126}{63}+\frac{7}{63}$$

$$=\frac{160}{63}$$

55.

$$\left.\begin{array}{l} 4 = 2\cdot 2 \\ 8 = 2\cdot 2\cdot 2 \end{array}\right\} LCD = 2\cdot 2\cdot 2 = 8$$

$$\frac{7}{8}-\frac{1}{4}=\frac{7}{8}-\frac{1\cdot 2}{4\cdot 2}$$

$$=\frac{7}{8}-\frac{2}{8}$$

$$=\frac{5}{8}$$

Applying the Concepts

57. Milk in carton 1 + Milk in carton 2 = total

$$\frac{1}{2}+4=\frac{1}{2}+\frac{4\cdot 2}{1\cdot 2}$$

$$=\frac{1}{2}+\frac{8}{2}$$

$$=\frac{9}{2}$$

The total is $\frac{9}{2}$ pints.

59. $\frac{5}{8}$ of monthly income = house payment

$$\frac{5}{8}\cdot 2{,}120=\frac{5\cdot 2{,}120}{8}$$

$$\frac{5\cdot \cancel{8}\cdot 265}{\cancel{8}}$$

$$=1{,}325$$

They can spend \$1,325 on a house payment.

61. Engineering fraction + business fraction = total fraction

$$\frac{1}{4}+\frac{3}{20}=\frac{5}{20}+\frac{3}{20}$$

$$=\frac{8}{20}$$

$$=\frac{2}{5}$$

63. The table is as follows:

Grade	Number	Fraction
A	$\frac{1}{8}\cdot 40 = 5$	$\frac{1}{8}$
B	$\frac{1}{5}\cdot 40 = 8$	$\frac{1}{5}$
C	$\frac{1}{2}\cdot 40 = 20$	$\frac{1}{2}$
below C	7	$\frac{7 \text{ students}}{40 \text{ total}}=\frac{7}{40}$
Total	40	1

65. Total acres ÷ acres per lot = lots

$$6 \div \frac{3}{5} = \frac{6}{1} \cdot \frac{5}{3}$$

$$= \frac{\cancel{3} \cdot 2 \cdot 5}{1 \cdot \cancel{3}}$$

$$= 10$$

There are 10 lots.

67. $P = 4s$

$$= 4 \cdot \left(\frac{3}{8} \text{ in} \right)$$

$$= \frac{4 \cdot 3}{8} \text{ in}$$

$$= \frac{\cancel{4} \cdot 3}{\cancel{4} \cdot 2} \text{ in}$$

$$= \frac{3}{2} \text{ in}$$

69. $P = $ sum of all sides

$$= \left(\frac{3}{5} + \frac{3}{5} + \frac{3}{10} + \frac{3}{10} \right) \text{ft}$$

$$= \left(\frac{3 \cdot 2}{5 \cdot 2} + \frac{3 \cdot 2}{5 \cdot 2} + \frac{3}{10} + \frac{3}{10} \right) \text{ft}$$

$$= \left(\frac{6}{10} + \frac{6}{10} + \frac{3}{10} + \frac{3}{10} \right) \text{ft}$$

$$= \frac{18}{10} \text{ft}$$

$$= \frac{9}{5} \text{ft}$$

Calculator Problems

71. Since $497 \div 71 = 7$, we have

$$\frac{7}{71} + \frac{17}{497} = \frac{7 \cdot 7}{71 \cdot 7} + \frac{17}{497}$$

$$= \frac{49}{497} + \frac{17}{497}$$

$$= \frac{66}{497}$$

73. Since
$1,247 \div 29 = 43$ and $1,247 \div 43 = 19$, we have:

$$\frac{10}{19} + \frac{20}{23} = \frac{10 \cdot 23}{19 \cdot 23} + \frac{20 \cdot 19}{23 \cdot 19}$$

$$= \frac{230}{437} + \frac{380}{437}$$

$$= \frac{610}{437}$$

Review Problems

75. $\dfrac{3}{4} \div \dfrac{5}{6} = \dfrac{3}{4} \cdot \dfrac{6}{5}$

$$= \frac{3 \cdot \cancel{2} \cdot 3}{\cancel{2} \cdot 2 \cdot 5}$$

$$= \frac{9}{10}$$

77. $12 \cdot \dfrac{2}{3} = \dfrac{12 \cdot 2}{3}$

$$= \frac{4 \cdot \cancel{3} \cdot 2}{\cancel{3}}$$

$$= 8$$

79. $4 \cdot \dfrac{3}{4} = \dfrac{\cancel{4} \cdot 3}{\cancel{4}} = 3$

81. $\dfrac{7}{6} \div \dfrac{7}{12} = \dfrac{7}{6} \cdot \dfrac{12}{7}$

$$= \frac{\cancel{7} \cdot 2 \cdot \cancel{6}}{\cancel{6} \cdot \cancel{7}}$$

$$= 2$$

83. $\dfrac{2}{3} \cdot \dfrac{3}{4} \cdot \dfrac{4}{5} \cdot \dfrac{5}{6} \cdot \dfrac{6}{7} = \dfrac{2 \cdot \cancel{3} \cdot \cancel{4} \cdot \cancel{5} \cdot \cancel{6}}{\cancel{3} \cdot \cancel{4} \cdot \cancel{5} \cdot \cancel{6} \cdot 7}$

$$= \frac{2}{7}$$

85. $\dfrac{35}{110} \cdot \dfrac{80}{63} \div \dfrac{16}{27} = \dfrac{35}{110} \cdot \dfrac{80}{63} \cdot \dfrac{27}{16}$

$$= \dfrac{\cancel{5} \cdot \cancel{7} \cdot \cancel{16} \cdot 5 \cdot \cancel{9} \cdot 3}{\cancel{5} \cdot 22 \cdot \cancel{9} \cdot \cancel{7} \cdot \cancel{16}}$$

$$= \dfrac{15}{22}$$

Extending the Concepts

87. $1, \dfrac{4}{3}, \dfrac{5}{3}, 2, \ldots$

The first term in the sequence is 1. Note:

$1 + \dfrac{1}{3} = \dfrac{4}{3}, \ \dfrac{4}{3} + \dfrac{1}{3} = \dfrac{5}{3}, \ \dfrac{5}{3} + \dfrac{1}{3} = \dfrac{6}{3} = 2,$ i.e. each term after 1 in the sequence comes

from adding $\dfrac{1}{3}$ to the previous term. The

next term in the sequence is

$$2 + \dfrac{1}{3} = \dfrac{2 \cdot 3}{1 \cdot 3} + \dfrac{1}{3} = \dfrac{7}{3}$$

89. $\dfrac{3}{2}, 2, \dfrac{5}{2}, \ldots$

The first term in the sequence is $\dfrac{1}{2}$. Note:

$\dfrac{3}{2} + \dfrac{1}{2} = \dfrac{4}{2} = 2$ and $\dfrac{4}{2} + \dfrac{1}{2} = \dfrac{5}{2}$, i.e. each term

after $\dfrac{3}{2}$ in the sequence comes from adding

$\dfrac{1}{2}$ to the previous term. The next term in

the sequence is $\dfrac{5}{2} + \dfrac{1}{2} = \dfrac{6}{2} = 3$

Section 2.6

1. $4\dfrac{2}{3} = \dfrac{3\cdot 4+2}{3} = \dfrac{12+2}{3} = \dfrac{14}{3}$

3. $5\dfrac{1}{4} = \dfrac{4\cdot 5+1}{4} = \dfrac{20+1}{4} = \dfrac{21}{4}$

5. $1\dfrac{5}{8} = \dfrac{8\cdot 1+5}{8} = \dfrac{8+5}{8} = \dfrac{13}{8}$

7. $15\dfrac{2}{3} = \dfrac{3\cdot 15+2}{3} = \dfrac{45+2}{3} = \dfrac{47}{3}$

9. $4\dfrac{20}{21} = \dfrac{21\cdot 4+20}{21} = \dfrac{84+20}{21} = \dfrac{104}{21}$

11. $12\dfrac{31}{33} = \dfrac{33\cdot 12+31}{33} = \dfrac{396+31}{33} = \dfrac{427}{33}$

13. $\dfrac{9}{8}:$ $8\overline{)9}$ so $\dfrac{9}{8} = 1+\dfrac{1}{8} = 1\dfrac{1}{8}$
$\dfrac{8}{1}$

15. $\dfrac{19}{4}:$ $4\overline{)19}$ so $\dfrac{19}{4} = 4+\dfrac{3}{4} = 4\dfrac{3}{4}$
$\dfrac{16}{3}$

17. $\dfrac{29}{6}:$ $6\overline{)29}$ so $\dfrac{29}{6} = 4+\dfrac{5}{6} = 4\dfrac{5}{6}$
$\dfrac{24}{5}$

19. $\dfrac{13}{4}:$ $4\overline{)13}$ so $\dfrac{13}{4} = 3+\dfrac{1}{4} = 3\dfrac{1}{4}$
$\dfrac{12}{1}$

21. $\dfrac{109}{27}:$ $27\overline{)109}$ so $\dfrac{109}{27} = 4+\dfrac{1}{27} = 4\dfrac{1}{27}$
$\dfrac{108}{1}$

23. $\dfrac{428}{15}:$ $15\overline{)428}$ so $\dfrac{428}{15} = 28+\dfrac{8}{15} = 28\dfrac{8}{15}$
$\dfrac{30\downarrow}{128}$
$\dfrac{120}{8}$

Applying the Concepts

25. a. $\dfrac{\text{Cicadas}}{\text{City traffic}} = \dfrac{90}{80} = \dfrac{9}{8} = 1\dfrac{1}{8}$

b. $\dfrac{\text{Firecracker}}{\text{City traffic}} = \dfrac{120}{80} = \dfrac{12}{8} = 1\dfrac{4}{8} = 1\dfrac{1}{2}$

27. Original price + increase = new price
$$5\dfrac{1}{4}+\dfrac{3}{4} = 5+\dfrac{1}{4}+\dfrac{3}{4}$$
$$= 5+\dfrac{1+3}{4}$$
$$= 5+\dfrac{4}{4}$$
$$= 5+1$$
$$= 6$$
The new price is $6 per share.

29. $5\dfrac{11}{12} = \dfrac{12\cdot 5+11}{12} = \dfrac{60+11}{12} = \dfrac{71}{12}$

The man is $\dfrac{71}{12}$ inches tall.

31. Above each age, draw a dot corresponding to the hours of sleep as measured by the vertical axis. To construct the line graph, connect the dots.

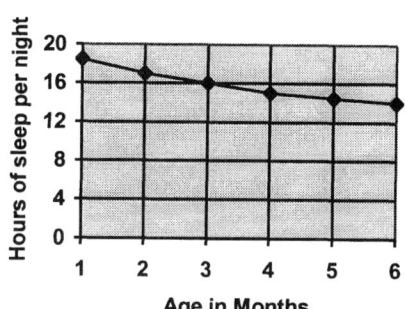

33. $135\dfrac{9}{10} = \dfrac{10 \cdot 135 + 9}{10} = \dfrac{1,350 + 9}{10} = \dfrac{1,359}{10}$

The price is $\dfrac{1,359}{10}$ ¢ per gallon.

Review Problems

35. $\dfrac{3}{8} \cdot \dfrac{3}{5} = \dfrac{3 \cdot 3}{8 \cdot 5} = \dfrac{9}{40}$

37. $\dfrac{2}{3}\left(\dfrac{9}{16}\right) = \dfrac{2 \cdot 9}{3 \cdot 16}$

$= \dfrac{\cancel{2} \cdot \cancel{3} \cdot 3}{\cancel{3} \cdot \cancel{2} \cdot 8}$

$= \dfrac{3}{8}$

39. $\dfrac{4}{5} \div \dfrac{7}{8} = \dfrac{4}{5} \cdot \dfrac{8}{7}$

$= \dfrac{4 \cdot 8}{5 \cdot 7}$

$= \dfrac{32}{35}$

41. $\dfrac{9}{10} \div \dfrac{3}{5} = \dfrac{9}{10} \cdot \dfrac{5}{3}$

$= \dfrac{3 \cdot \cancel{3} \cdot \cancel{5}}{\cancel{5} \cdot 2 \cdot \cancel{3}}$

$= \dfrac{3}{2}$ or $1\dfrac{1}{2}$

43. $\dfrac{2}{3} \cdot \dfrac{3}{4} \div \dfrac{5}{8} = \dfrac{2}{3} \cdot \dfrac{3}{4} \cdot \dfrac{8}{5}$

$= \dfrac{2 \cdot \cancel{3} \cdot 2 \cdot \cancel{4}}{\cancel{3} \cdot \cancel{4} \cdot 5}$

$= \dfrac{4}{5}$

Improving your Quantitative Literacy

45. a. $1\dfrac{1}{2} \cdot 1,405,146 = \dfrac{3}{2} \cdot \dfrac{1,405,146}{1}$

$= \dfrac{3 \cdot \cancel{2} \cdot 702,573}{\cancel{2}}$

$= 2,107,719$

b. $1\dfrac{1}{3} \cdot 1,405,146 = \dfrac{4}{3} \cdot \dfrac{1,405,146}{1}$

$= \dfrac{4 \cdot \cancel{3} \cdot 468,382}{\cancel{3}}$

$= 1,873,528$

(a) is closer to the truth; 2,107,719 is closer to 2,082,487 than 1,873,528 is.

Section 2.7

1. $3\dfrac{2}{5} \cdot 1\dfrac{1}{2} = \dfrac{17}{5} \cdot \dfrac{3}{2}$

$\qquad = \dfrac{51}{10}$

$\qquad = 5\dfrac{1}{10}$

3. $5\dfrac{1}{8} \cdot 2\dfrac{2}{3} = \dfrac{41}{8} \cdot \dfrac{8}{3}$

$\qquad = \dfrac{41 \cdot \cancel{8}}{\cancel{8} \cdot 3}$

$\qquad = \dfrac{41}{3}$

$\qquad = 13\dfrac{2}{3}$

5. $2\dfrac{1}{10} \cdot 3\dfrac{3}{10} = \dfrac{21}{10} \cdot \dfrac{33}{10}$

$\qquad = \dfrac{693}{100}$

$\qquad = 6\dfrac{93}{100}$

7. $1\dfrac{1}{4} \cdot 4\dfrac{2}{3} = \dfrac{5}{4} \cdot \dfrac{14}{3}$

$\qquad = \dfrac{5 \cdot \cancel{2} \cdot 7}{\cancel{2} \cdot 2 \cdot 3}$

$\qquad = \dfrac{35}{6}$

$\qquad = 5\dfrac{5}{6}$

9. $2 \cdot 4\dfrac{7}{8} = \dfrac{2}{1} \cdot \dfrac{39}{8}$

$\qquad = \dfrac{\cancel{2} \cdot 39}{1 \cdot \cancel{2} \cdot 4}$

$\qquad = \dfrac{39}{4}$

$\qquad = 9\dfrac{3}{4}$

11. $\dfrac{3}{5} \cdot 5\dfrac{1}{3} = \dfrac{3}{5} \cdot \dfrac{16}{3}$

$\qquad = \dfrac{\cancel{3} \cdot 16}{5 \cdot \cancel{3}}$

$\qquad = \dfrac{16}{5}$

$\qquad = 3\dfrac{1}{5}$

13. $2\dfrac{1}{2} \cdot 3\dfrac{1}{3} \cdot 1\dfrac{1}{2} = \dfrac{5}{2} \cdot \dfrac{10}{3} \cdot \dfrac{3}{2}$

$\qquad = \dfrac{5 \cdot \cancel{2} \cdot 5 \cdot \cancel{3}}{\cancel{2} \cdot \cancel{3} \cdot 2}$

$\qquad = \dfrac{25}{2}$

$\qquad = 12\dfrac{1}{2}$

15. $\dfrac{3}{4} \cdot 7 \cdot 1\dfrac{4}{5} = \dfrac{3}{4} \cdot \dfrac{7}{1} \cdot \dfrac{9}{5}$

$\qquad = \dfrac{3 \cdot 7 \cdot 9}{4 \cdot 1 \cdot 5}$

$\qquad = \dfrac{189}{20}$

$\qquad = 9\dfrac{9}{20}$

17. $3\dfrac{1}{5} \div 4\dfrac{1}{2} = \dfrac{16}{5} \div \dfrac{9}{2}$

$\qquad = \dfrac{16}{5} \cdot \dfrac{2}{9}$

$\qquad = \dfrac{32}{45}$

19. $6\dfrac{1}{4} \div 3\dfrac{3}{4} = \dfrac{25}{4} \div \dfrac{15}{4}$

$\qquad = \dfrac{25}{4} \cdot \dfrac{4}{15}$

$\qquad = \dfrac{\cancel{5} \cdot 5 \cdot \cancel{4}}{\cancel{4} \cdot \cancel{5} \cdot 3}$

$\qquad = \dfrac{5}{3}$

$\qquad = 1\dfrac{2}{3}$

21. $10 \div 2\dfrac{1}{2} = \dfrac{10}{1} \div \dfrac{5}{2}$

$\qquad = \dfrac{10}{1} \cdot \dfrac{2}{5}$

$\qquad = \dfrac{2 \cdot \cancel{5} \cdot 2}{1 \cdot \cancel{5}}$

$\qquad = 4$

23. $8\dfrac{3}{5} \div 2 = \dfrac{43}{5} \div \dfrac{2}{1}$

$\qquad = \dfrac{43}{5} \cdot \dfrac{1}{2}$

$\qquad = \dfrac{43 \cdot 1}{5 \cdot 2}$

$\qquad = \dfrac{43}{10}$

$\qquad = 4\dfrac{3}{10}$

25. $\left(\dfrac{3}{4} \div 2\dfrac{1}{2}\right) \div 3 = \left(\dfrac{3}{4} \div \dfrac{5}{2}\right) \div \dfrac{3}{1}$

$\qquad = \left(\dfrac{3}{4} \cdot \dfrac{2}{5}\right) \cdot \dfrac{1}{3}$

$\qquad = \dfrac{\cancel{3} \cdot \cancel{2} \cdot 1}{\cancel{2} \cdot 2 \cdot 5 \cdot \cancel{3}}$

$\qquad = \dfrac{1}{10}$

27. $\left(8 \div 1\dfrac{1}{4}\right) \div 2 = \left(\dfrac{8}{1} \div \dfrac{5}{4}\right) \div \dfrac{2}{1}$

$\qquad = \left(\dfrac{8}{1} \cdot \dfrac{4}{5}\right) \cdot \dfrac{1}{2}$

$\qquad = \dfrac{8 \cdot 2 \cdot \cancel{2} \cdot 1}{1 \cdot 5 \cdot \cancel{2}}$

$\qquad = \dfrac{16}{5}$

$\qquad = 3\dfrac{1}{5}$

29. $2\dfrac{1}{2} \cdot \left(3\dfrac{2}{5} \div 4\right) = \dfrac{5}{2} \cdot \left(\dfrac{17}{5} \div \dfrac{4}{1}\right)$

$\qquad = \dfrac{5}{2} \cdot \left(\dfrac{17}{5} \cdot \dfrac{1}{4}\right)$

$\qquad = \dfrac{\cancel{5} \cdot 17 \cdot 1}{2 \cdot \cancel{5} \cdot 4}$

$\qquad = \dfrac{17}{8}$

$\qquad = 2\dfrac{1}{8}$

31. $2\dfrac{1}{2} \cdot 3 = \dfrac{5}{2} \cdot \dfrac{3}{1}$

$\qquad = \dfrac{5 \cdot 3}{2 \cdot 1}$

$\qquad = \dfrac{15}{2}$

$\qquad = 7\dfrac{1}{2}$

33. $2\dfrac{3}{4} \div 3\dfrac{1}{4} = \dfrac{11}{4} \div \dfrac{13}{4}$

$\qquad = \dfrac{11}{4} \cdot \dfrac{4}{13}$

$\qquad = \dfrac{11 \cdot \cancel{4}}{\cancel{4} \cdot 13}$

$\qquad = \dfrac{11}{13}$

Applying the Concepts

35. $2 \cdot \left(2\dfrac{3}{4} \right) = \dfrac{2}{1} \cdot \dfrac{11}{4}$

$\qquad = \dfrac{\cancel{2} \cdot 11}{1 \cdot \cancel{2} \cdot 2}$

$\qquad = \dfrac{11}{2}$

$\qquad = 5\dfrac{1}{2}$

$5\dfrac{1}{2}$ cups will be needed.

37. $\dfrac{1}{3}$ of $2\dfrac{1}{2}$ cups = amount needed

$\dfrac{1}{3} \cdot 2\dfrac{1}{2} = \dfrac{1}{3} \cdot \dfrac{5}{2}$

$\qquad = \dfrac{1 \cdot 5}{3 \cdot 2}$

$\qquad = \dfrac{5}{6}$

$\dfrac{5}{6}$ cup will be needed.

39. $\dfrac{3}{4}$ of $1\dfrac{7}{9} = \dfrac{3}{4} \cdot \dfrac{16}{9}$

$\qquad = \dfrac{\cancel{3} \cdot \cancel{4} \cdot 4}{\cancel{4} \cdot \cancel{3} \cdot 3}$

$\qquad = \dfrac{4}{3}$

$\qquad = 1\dfrac{1}{3}$

41. 8 gallons × cost per gallon = total cost

$8 \cdot 135\dfrac{9}{10} = \dfrac{8}{1} \cdot \dfrac{1,359}{10}$

$\qquad = \dfrac{\cancel{2} \cdot 4 \cdot 1,359}{1 \cdot \cancel{2} \cdot 5}$

$\qquad = \dfrac{5,436}{5}$

$\qquad = 1,087\dfrac{1}{5}$

The total cost is $1,087\dfrac{1}{5}$¢.

43. 5 gallons × miles per gallon = total miles

$5 \cdot 32\dfrac{3}{4} = \dfrac{5}{1} \cdot \dfrac{131}{4}$

$\qquad = \dfrac{5 \cdot 131}{1 \cdot 4}$

$\qquad = \dfrac{655}{4}$

$\qquad = 163\dfrac{3}{4}$

The car will travel $163\dfrac{3}{4}$ miles.

45. 3 covers × yards per cover = total yards

$$3 \cdot 1\frac{1}{2} = \frac{3}{1} \cdot \frac{3}{2}$$

$$= \frac{3 \cdot 3}{1 \cdot 2}$$

$$= \frac{9}{2}$$

$$= 4\frac{1}{2}$$

It will take $4\frac{1}{2}$ yards.

47. $A = \frac{1}{2}bh$

$$= \frac{1}{2} \cdot 3 \text{ yd} \cdot 1\frac{1}{2} \text{ yd}$$

$$= \frac{1}{2} \cdot \frac{3}{1} \cdot \frac{3}{2} \text{ yd}^2$$

$$= \frac{9}{4} \text{ yd}^2$$

$$= 2\frac{1}{4} \text{ yd}^2$$

49.

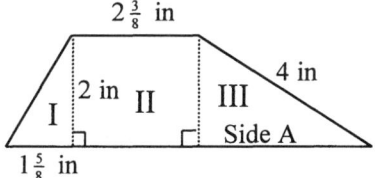

Separate into 3 regions. Find the length of the missing side:

$$\text{Side A} = 7 - 1\frac{5}{8} - 2\frac{3}{8}$$

$$= \frac{7}{1} - \frac{13}{8} - \frac{19}{8}$$

$$= \frac{7 \cdot 8}{1 \cdot 8} - \frac{13}{8} - \frac{19}{8}$$

$$= \frac{56}{8} - \frac{13}{8} - \frac{19}{8}$$

$$= \frac{24}{8} = 3 \text{ in.}$$

49. (continued)

$$\text{Area I} = \frac{1}{2}bh \qquad \text{Area II} = lw$$

$$= \frac{1}{2}\left(1\frac{5}{8}\right) \cdot 2 \qquad = 2\frac{3}{8} \cdot 2$$

$$= \frac{1}{2} \cdot \frac{13}{8} \cdot \frac{2}{1} \qquad = \frac{19}{8} \cdot \frac{2}{1}$$

$$= \frac{13 \cdot \cancel{2}}{\cancel{2} \cdot 8 \cdot 1} \qquad = \frac{19 \cdot \cancel{2}}{\cancel{2} \cdot 4 \cdot 1}$$

$$= \frac{13}{8} \qquad\qquad = \frac{19}{4}$$

$$= 1\frac{5}{8} \text{ in}^2 \qquad = 4\frac{3}{4} \text{ in}^2$$

$$\text{Area III} = \frac{1}{2}bh = \frac{1}{2} \cdot 3 \cdot 2$$

$$= \frac{1}{2} \cdot \frac{3}{1} \cdot \frac{\cancel{2}}{1}$$

$$= 3 \text{ in}^2$$

Area = Area I + Area II + Area III

$$A = 1\frac{5}{8} + 4\frac{3}{4} + 3$$

$$= \frac{13}{8} + \frac{19}{4} + \frac{3}{1}$$

$$= \frac{13}{8} + \frac{19 \cdot 2}{4 \cdot 2} + \frac{3 \cdot 8}{1 \cdot 8}$$

$$= \frac{13 + 38 + 24}{8}$$

$$= \frac{75}{8} \text{ or } 9\frac{3}{8} \text{ in}^2$$

51. Can 1 calories − Can 2 calories = difference

$$45 \cdot 3\frac{1}{2} - 25 \cdot 3\frac{1}{2} = (45 - 25) \cdot 3\frac{1}{2}$$

$$= 20 \cdot 3\frac{1}{2}$$

$$= \frac{20}{1} \cdot \frac{7}{2}$$

$$= \frac{140}{2}$$

$$= 70$$

Can 1 has 70 more calories than can 2.

53. Can 1 sodium − Can 2 sodium = difference

$$560 \cdot 3\frac{1}{2} - 300 \cdot 3\frac{1}{2} = (560 - 300) \cdot 3\frac{1}{2}$$

$$= 260 \cdot 3\frac{1}{2}$$

$$= \frac{260}{1} \cdot \frac{7}{2}$$

$$= \frac{1,820}{2}$$

$$= 910$$

Can 1 has 910 mg more sodium than Can 2.

Review Problems

55. $\dfrac{3}{4} + \dfrac{5}{4} = \dfrac{3+5}{4} = \dfrac{8}{4} = 2$

57.
$$\left. \begin{array}{l} 56 = 2 \cdot 2 \cdot 2 \cdot 7 \\ 42 = 2 \cdot 3 \cdot 7 \end{array} \right\} \text{LCD} = 2 \cdot 2 \cdot 2 \cdot 3 \cdot 7 = 168$$

$$\frac{15}{56} - \frac{5}{42} = \frac{15 \cdot 3}{56 \cdot 3} - \frac{5 \cdot 4}{42 \cdot 4}$$

$$= \frac{45}{168} - \frac{20}{168}$$

$$= \frac{25}{168}$$

59.
$$\left. \begin{array}{l} 18 = 2 \cdot 3 \cdot 3 \\ 12 = 2 \cdot 2 \cdot 3 \\ 4 = 2 \cdot 2 \end{array} \right\} \text{LCD} = 2 \cdot 2 \cdot 3 \cdot 3 = 36$$

$$\frac{5}{18} + \frac{1}{12} + \frac{3}{4} = \frac{5 \cdot 2}{18 \cdot 2} + \frac{1 \cdot 3}{12 \cdot 3} + \frac{3 \cdot 9}{4 \cdot 9}$$

$$= \frac{10}{36} + \frac{3}{36} + \frac{27}{36}$$

$$= \frac{40}{36}$$

$$= \frac{10}{9}$$

$$= 1\frac{1}{9}$$

Extending the Concepts

61. $\left(1\dfrac{1}{2}\right)^2 = \left(\dfrac{3}{2}\right)^2 = \dfrac{3}{2} \cdot \dfrac{3}{2} = \dfrac{9}{4} = 2\dfrac{1}{4}$

63. $\left(1\dfrac{3}{4}\right)^2 = \left(\dfrac{7}{4}\right)^2 = \dfrac{7}{4} \cdot \dfrac{7}{4} = \dfrac{49}{16} = 3\dfrac{1}{16}$

65. $A = 2\dfrac{1}{4}$ ft^2 $= \dfrac{9}{4}$ ft^2.

Because $A = s^2$, we need to find s so that

$s^2 = s \cdot s = \dfrac{9}{4}$ ft^2. By inspection,

$s = \dfrac{3}{2}$ ft $= 1\dfrac{1}{2}$ ft.

67. $V = 3\dfrac{3}{8}$ ft^3 $= \dfrac{27}{8}$ ft^3.

Because $V = s^3$, we need to find s so that

$s^3 = s \cdot s \cdot s = \dfrac{27}{8}$ ft^3. By inspection,

$s = \dfrac{3}{2}$ ft $= 1\dfrac{1}{2}$ ft.

Section 2.8

1.
$$2\frac{1}{5}$$
$$+\ 3\frac{3}{5}$$
$$5\frac{4}{5}$$

3.
$$4\frac{3}{10}$$
$$+\ 8\frac{1}{10}$$
$$12\frac{4}{10} = 12\frac{2}{5}$$

5.
$$6\frac{8}{9}$$
$$-\ 3\frac{4}{9}$$
$$3\frac{4}{9}$$

7.
$$9\frac{1}{6}$$
$$+\ 2\frac{5}{6}$$
$$11\frac{6}{6} = 11 + 1 = 12$$

9.
$$3\frac{5}{8} = \quad 3\frac{5}{8} \quad = \quad 3\frac{5}{8}$$
$$-\ 2\frac{1}{4} = -\ 2\frac{1\cdot 2}{4\cdot 2} = -\ 2\frac{2}{8}$$
$$1\frac{3}{8}$$

11.
$$11\frac{1}{3} = \quad 11\frac{1\cdot 2}{3\cdot 2} = \quad 11\frac{2}{6}$$
$$+\ 2\frac{5}{6} = +\ 2\frac{5}{6} \quad = +\ 2\frac{5}{6}$$
$$13\frac{7}{6} = 14\frac{1}{6}$$

13.
$$7\frac{5}{12} = \quad 7\frac{5}{12} = \quad 7\frac{5}{12}$$
$$-\ 3\frac{1}{3} = -\ 3\frac{1\cdot 4}{3\cdot 4} = -\ 3\frac{4}{12}$$
$$4\frac{1}{12}$$

15.
$$6\frac{1}{3} = \quad 6\frac{1\cdot 4}{3\cdot 4} = \quad 6\frac{4}{12}$$
$$-\ 4\frac{1}{4} = -\ 4\frac{1\cdot 3}{4\cdot 3} = -\ 4\frac{3}{12}$$
$$2\frac{1}{12}$$

17.
$$10\frac{5}{6} = \quad 10\frac{5\cdot 2}{6\cdot 2} = \quad 10\frac{10}{12}$$
$$+\ 15\frac{3}{4} = +\ 15\frac{3\cdot 3}{4\cdot 3} = +\ 15\frac{9}{12}$$
$$25\frac{19}{12} = 26\frac{7}{12}$$

19.
$$5\frac{2}{3}$$
$$+\ 6\frac{1}{3}$$
$$11\frac{3}{3} = 12$$

21.
$$10\frac{13}{16}$$
$$-\ 8\frac{5}{16}$$
$$2\frac{8}{16} = 2\frac{1}{2}$$

23.
$$6\frac{1}{2} = 6\frac{1\cdot 7}{2\cdot 7} = 6\frac{7}{14}$$
$$+\ 2\frac{5}{14} = +\ 2\frac{5}{14} = +\ 2\frac{5}{14}$$
$$8\frac{12}{14} = 8\frac{6}{7}$$

25.
$$1\frac{5}{8} = 1\frac{5}{8} = 1\frac{5}{8}$$
$$+\ 1\frac{3}{4} = +\ 1\frac{3\cdot 2}{4\cdot 2} = +\ 1\frac{6}{8}$$
$$2\frac{11}{8} = 3\frac{3}{8}$$

27.
$$4\frac{2}{3} = 4\frac{2\cdot 5}{3\cdot 5} = 4\frac{10}{15}$$
$$+\ 5\frac{3}{5} = +\ 5\frac{3\cdot 3}{5\cdot 3} = +\ 5\frac{9}{15}$$
$$9\frac{19}{15} = 10\frac{4}{15}$$

29.
$$5\frac{4}{10} = 5\frac{4\cdot 3}{10\cdot 3} = 5\frac{12}{30}$$
$$-\ 3\frac{1}{3} = -\ 3\frac{1\cdot 10}{3\cdot 10} = -\ 3\frac{10}{30}$$
$$2\frac{2}{30} = 2\frac{1}{15}$$

31.
$$1\frac{1}{4}$$
$$2\frac{3}{4}$$
$$+\ 5$$
$$8\frac{4}{4} = 9$$

33.
$$7\frac{1}{10}$$
$$8\frac{3}{10}$$
$$+\ 2\frac{7}{10}$$
$$17\frac{11}{10} = 18\frac{1}{10}$$

35.
$$\frac{3}{4}$$
$$8\frac{1}{4}$$
$$+\ 5$$
$$13\frac{4}{4} = 14$$

37.
$$3\frac{1}{2} = 3\frac{1\cdot 3}{2\cdot 3} = 3\frac{3}{6}$$
$$8\frac{1}{3} = 8\frac{1\cdot 2}{3\cdot 2} = 8\frac{2}{6}$$
$$+\ 5\frac{1}{6} = +\ 5\frac{1}{6} = +\ 5\frac{1}{6}$$
$$16\frac{6}{6} = 17$$

39.
$$8\frac{2}{3} = 8\frac{2\cdot8}{3\cdot8} = 8\frac{16}{24}$$
$$9\frac{1}{8} = 9\frac{1\cdot3}{8\cdot3} = 9\frac{3}{24}$$
$$+\,6\frac{1}{4} = +\,6\frac{1\cdot6}{4\cdot6} = +\,6\frac{6}{24}$$
$$23\frac{25}{24} = 24\frac{1}{24}$$

41.
$$6\frac{1}{7} = 6\frac{1\cdot2}{7\cdot2} = 6\frac{2}{14}$$
$$9\frac{3}{14} = 9\frac{3}{14} = 9\frac{3}{14}$$
$$+\,12\frac{1}{2} = +\,12\frac{1\cdot7}{2\cdot7} = +\,12\frac{7}{14}$$
$$27\frac{12}{14} = 27\frac{6}{7}$$

43.
$$10\frac{1}{20} = 10\frac{1}{20} = 10\frac{1}{20}$$
$$11\frac{4}{5} = 11\frac{4\cdot4}{5\cdot4} = 11\frac{16}{20}$$
$$+\,15\frac{3}{10} = +\,15\frac{3\cdot2}{10\cdot2} = +\,15\frac{6}{20}$$
$$36\frac{23}{20} = 37\frac{3}{20}$$

45.
$$8 = 7\frac{4}{4}$$
$$-\,1\frac{3}{4} = -\,1\frac{3}{4}$$
$$6\frac{1}{4}$$

47.
$$15 = 14\frac{10}{10}$$
$$-\,5\frac{3}{10} = -\,5\frac{3}{10}$$
$$9\frac{7}{10}$$

49.
$$8\frac{1}{4} = 7\frac{5}{4}$$
$$-\,2\frac{3}{4} = -\,2\frac{3}{4}$$
$$5\frac{2}{4} = 5\frac{1}{2}$$

51.
$$9\frac{1}{3} = 8\frac{4}{3}$$
$$-\,8\frac{2}{3} = -\,8\frac{2}{3}$$
$$\frac{2}{3}$$

53.
$$4\frac{1}{4} = 4\frac{1\cdot3}{4\cdot3} = 4\frac{3}{12} = 3\frac{15}{12}$$
$$-\,2\frac{1}{3} = -\,2\frac{1\cdot4}{3\cdot4} = -\,2\frac{4}{12} = -\,2\frac{4}{12}$$
$$1\frac{11}{12}$$

55.
$$9\frac{2}{3} = 9\frac{2\cdot4}{3\cdot4} = 9\frac{8}{12} = 8\frac{20}{12}$$
$$-\,5\frac{3}{4} = -\,5\frac{3\cdot3}{4\cdot3} = -\,5\frac{9}{12} = -\,5\frac{9}{12}$$
$$3\frac{11}{12}$$

57.
$$16\frac{3}{4} = 16\frac{3\cdot5}{4\cdot5} = 16\frac{15}{20} = 15\frac{35}{20}$$
$$-\,10\frac{4}{5} = -\,10\frac{4\cdot4}{5\cdot4} = -\,10\frac{16}{20} = -\,10\frac{16}{20}$$
$$5\frac{19}{20}$$

59.
$$10\frac{3}{10} = 10\frac{3}{10} = 10\frac{3}{10} = 9\frac{13}{10}$$
$$-4\frac{4}{5} = -4\frac{4\cdot2}{5\cdot2} = -4\frac{8}{10} = -4\frac{8}{10}$$
$$5\frac{5}{10} = 5\frac{1}{2}$$

61.
$$13\frac{1}{6} = 13\frac{1\cdot4}{6\cdot4} = 13\frac{4}{24} = 12\frac{28}{24}$$
$$-12\frac{5}{8} = -.12\frac{5\cdot3}{8\cdot3} = -12\frac{15}{24} = -12\frac{15}{24}$$
$$\frac{13}{24}$$

63.
$$6\frac{1}{5} = 6\frac{1\cdot2}{5\cdot2} = 6\frac{2}{10} = 5\frac{12}{10}$$
$$-2\frac{7}{10} = -2\frac{7}{10} = -2\frac{7}{10} = -2\frac{7}{10}$$
$$3\frac{5}{10} = 3\frac{1}{2}$$

65.
$$3\frac{1}{8} = 3\frac{1\cdot5}{8\cdot5} = 3\frac{5}{40}$$
$$+2\frac{3}{5} = +2\frac{3\cdot8}{5\cdot8} = +2\frac{24}{40}$$
$$5\frac{29}{40}$$

Applying the Concepts

67. Add the two lengths:
$$5\frac{7}{8}$$
$$+6\frac{3}{8}$$
$$11\frac{10}{8} = 12\frac{2}{8} = 12\frac{1}{4}$$

The total length is $12\frac{1}{4}$ inches.

69. Find the difference between the distances:
$$1\frac{1}{2} = 1\frac{1\cdot4}{2\cdot4} = 1\frac{4}{8}$$
$$-1\frac{3}{8} = -1\frac{3}{8} = -1\frac{3}{8}$$
$$\frac{1}{8}$$

The horse ran $\frac{1}{8}$ mile further.

71. Original length – shrinkage = new length
$$32\frac{1}{2} = 32\frac{1\cdot3}{2\cdot3} = 32\frac{3}{6}$$
$$-1\frac{1}{3} = -1\frac{1\cdot2}{3\cdot2} = -1\frac{2}{6}$$
$$31\frac{1}{6}$$

The new length is $31\frac{1}{6}$ inches.

73. NFL: $P = 2l + 2w$

$$= 2 \cdot 100 \text{ yd} + 2 \cdot 53\frac{1}{3} \text{ yd}$$

$$= \frac{200}{1} \text{ yd} + \frac{2}{1} \cdot \frac{160}{3} \text{ yd}$$

$$= \frac{600}{3} \text{ yd} + \frac{320}{3} \text{ yd}$$

$$= \frac{920}{3} \text{ yd} = 306\frac{2}{3} \text{ yd}$$

Canadian: $P = 2l + 2w$

$$= 2 \cdot 110 \text{ yd} + 2 \cdot 65 \text{ yd}$$

$$= 220 \text{ yd} + 130 \text{ yd}$$

$$= 350 \text{ yd}$$

Arena: $P = 2l + 2w$

$$= 2 \cdot 50 \text{ yd} + 2 \cdot 28\frac{1}{3} \text{ yd}$$

$$= \frac{100}{1} \text{ yd} + \frac{2}{1} \cdot \frac{85}{3} \text{ yd}$$

$$= \frac{300}{3} \text{ yd} + \frac{170}{3} \text{ yd}$$

$$= \frac{470}{3} \text{ yd} = 156\frac{2}{3} \text{ yd}$$

75.   a.   3/13 price − 3/8 price = difference

$$
\begin{array}{rcl}
77\dfrac{3}{8} & = & 76\dfrac{11}{8} \\[2mm]
-\ 74\dfrac{7}{8} & = & -\ 74\dfrac{7}{8} \\[2mm]
\hline
& & 2\dfrac{4}{8} = 2\dfrac{1}{2}
\end{array}
$$

The difference is $\$2\frac{1}{2}$.

   b.   100 shares × difference in price = net gain

$$100 \cdot \$2\frac{1}{2} = 100 \cdot \frac{5}{2} = \$250$$

The stock is worth \$250 more.

77.   200 shares × difference in price = net gain/loss

$$200\left(34\frac{3}{8} - 32\frac{7}{8}\right) = 200\left(33\frac{11}{8} - 32\frac{7}{8}\right)$$

$$= 200\left(1\frac{4}{8}\right)$$

$$= 200\left(1\frac{1}{2}\right)$$

$$= 200 \cdot \frac{3}{2}$$

$$= 300$$

The stock would be worth \$300 more.

Review Problems

79. $3 + 2 \cdot 7 = 3 + 14$

$$= 17$$

81. $4 \cdot 5 - 3 \cdot 2 = 20 - 6$

$$= 14$$

83. $3 \cdot 2^3 + 5 \cdot 4^2 = 3 \cdot 8 + 5 \cdot 16$

$$= 24 + 80$$

$$= 104$$

85. $3\left[2 + 5(6)\right] = 3\left[2 + 30\right]$

$$= 3\left[32\right]$$

$$= 96$$

87. $(7 - 3)(8 + 2) = (4)(10)$

$$= 40$$

Improving Your Quantitative Literacy

89.   Halfbridle's speed:

$$
\begin{array}{rcl}
35\dfrac{1}{5} & = & 34\dfrac{6}{5} \\[2mm]
-\ \dfrac{2}{5} & = & -\ \dfrac{2}{5} \\[2mm]
\hline
& & 34\dfrac{4}{5} \text{ sec}
\end{array}
$$

Section 2.9

1. $3 + \left(1\frac{1}{2}\right)\left(2\frac{2}{3}\right) = 3 + \left(\frac{3}{2}\right)\left(\frac{8}{3}\right)$

$\qquad\qquad\qquad = 3 + 4$

$\qquad\qquad\qquad = 7$

3. $8 - \left(\frac{6}{11}\right)\left(1\frac{5}{6}\right) = 8 - \left(\frac{6}{11}\right)\left(\frac{11}{6}\right)$

$\qquad\qquad\qquad = 8 - 1$

$\qquad\qquad\qquad = 7$

5. $\frac{2}{3}\left(1\frac{1}{2}\right) + \frac{3}{4}\left(1\frac{1}{3}\right) = \left(\frac{2}{3}\right)\left(\frac{3}{2}\right) + \left(\frac{3}{4}\right)\left(\frac{4}{3}\right)$

$\qquad\qquad\qquad = 1 + 1$

$\qquad\qquad\qquad = 2$

7. $2\left(1\frac{1}{2}\right) + 5\left(6\frac{2}{5}\right) = 2\left(\frac{3}{2}\right) + 5\left(\frac{32}{5}\right)$

$\qquad\qquad\qquad = 3 + 32$

$\qquad\qquad\qquad = 35$

9. $\left(\frac{3}{5} + \frac{1}{10}\right)\left(\frac{1}{2} + \frac{3}{4}\right) = \left(\frac{3 \cdot 2}{5 \cdot 2} + \frac{1}{10}\right)\left(\frac{1 \cdot 2}{2 \cdot 2} + \frac{3}{4}\right)$

$\qquad\qquad\qquad = \left(\frac{6}{10} + \frac{1}{10}\right)\left(\frac{2}{4} + \frac{3}{4}\right)$

$\qquad\qquad\qquad = \left(\frac{7}{10}\right)\left(\frac{5}{4}\right)$

$\qquad\qquad\qquad = \frac{7}{8}$

11. $\left(2 + \frac{2}{3}\right)\left(3 + \frac{1}{8}\right) = \left(2\frac{2}{3}\right)\left(3\frac{1}{8}\right)$

$\qquad\qquad\qquad = \left(\frac{8}{3}\right)\left(\frac{25}{8}\right)$

$\qquad\qquad\qquad = \frac{25}{3}$

$\qquad\qquad\qquad = 8\frac{1}{3}$

13. $\left(1 + \frac{5}{6}\right)\left(1 - \frac{5}{6}\right) = \left(\frac{6}{6} + \frac{5}{6}\right)\left(\frac{6}{6} - \frac{5}{6}\right)$

$\qquad\qquad\qquad = \left(\frac{11}{6}\right)\left(\frac{1}{6}\right)$

$\qquad\qquad\qquad = \frac{11}{36}$

15. $\frac{2}{3} + \frac{1}{3}\left(2\frac{1}{2} + \frac{1}{2}\right)^2 = \frac{2}{3} + \frac{1}{3}(3)^2$

$\qquad\qquad\qquad = \frac{2}{3} + \frac{1}{3}(9)$

$\qquad\qquad\qquad = \frac{2}{3} + 3$

$\qquad\qquad\qquad = 3\frac{2}{3}$

17. $2\frac{3}{8} + \frac{1}{2}\left(\frac{1}{3} + \frac{5}{3}\right)^3 = 2\frac{3}{8} + \frac{1}{2}\left(\frac{6}{3}\right)^3$

$\qquad\qquad\qquad = 2\frac{3}{8} + \frac{1}{2}(2)^3$

$\qquad\qquad\qquad = 2\frac{3}{8} + \frac{1}{2}(8)$

$\qquad\qquad\qquad = 2\frac{3}{8} + 4$

$\qquad\qquad\qquad = 6\frac{3}{8}$

19. $2\left(\dfrac{1}{2}+\dfrac{1}{3}\right)+3\left(\dfrac{2}{3}+\dfrac{1}{4}\right)$

$=2\cdot\dfrac{1}{2}+2\cdot\dfrac{1}{3}+3\cdot\dfrac{2}{3}+3\cdot\dfrac{1}{4}$

$=1+\dfrac{2}{3}+2+\dfrac{3}{4}$

$=(1+2)+\left(\dfrac{2\cdot4}{3\cdot4}+\dfrac{3\cdot3}{4\cdot3}\right)$

$=3+\left(\dfrac{8}{12}+\dfrac{9}{12}\right)$

$=3\dfrac{17}{12}$

$=4\dfrac{5}{12}$

21. $\dfrac{\dfrac{2}{3}}{\dfrac{3}{4}}=\dfrac{2}{3}\div\dfrac{3}{4}$

$\quad=\dfrac{2}{3}\cdot\dfrac{4}{3}$

$\quad=\dfrac{8}{9}$

23. $\dfrac{\dfrac{2}{3}}{\dfrac{4}{3}}=\dfrac{2}{3}\div\dfrac{4}{3}$

$\quad=\dfrac{2}{3}\cdot\dfrac{3}{4}$

$\quad=\dfrac{2}{4}$

$\quad=\dfrac{1}{2}$

25. $\dfrac{\dfrac{11}{20}}{\dfrac{5}{10}}=\dfrac{11}{20}\div\dfrac{5}{10}$

$\quad=\dfrac{11}{20}\cdot\dfrac{10}{5}$

$\quad=\dfrac{11}{10}$

$\quad=1\dfrac{1}{10}$

27. $\dfrac{\dfrac{1}{2}+\dfrac{1}{3}}{\dfrac{1}{2}-\dfrac{1}{3}}=\dfrac{6\left(\dfrac{1}{2}+\dfrac{1}{3}\right)}{6\left(\dfrac{1}{2}-\dfrac{1}{3}\right)}$

$\quad=\dfrac{6\cdot\dfrac{1}{2}+6\cdot\dfrac{1}{3}}{6\cdot\dfrac{1}{2}-6\cdot\dfrac{1}{3}}$

$\quad=\dfrac{3+2}{3-2}$

$\quad=5$

29. $\dfrac{\dfrac{5}{8}-\dfrac{1}{4}}{\dfrac{1}{8}+\dfrac{1}{2}}=\dfrac{8\left(\dfrac{5}{8}-\dfrac{1}{4}\right)}{8\left(\dfrac{1}{8}+\dfrac{1}{2}\right)}$

$\quad=\dfrac{8\cdot\dfrac{5}{8}-8\cdot\dfrac{1}{4}}{8\cdot\dfrac{1}{8}+8\cdot\dfrac{1}{2}}$

$\quad=\dfrac{5-2}{1+4}$

$\quad=\dfrac{3}{5}$

31. $\dfrac{\dfrac{9}{20}-\dfrac{1}{10}}{\dfrac{1}{10}+\dfrac{9}{20}}=\dfrac{20\left(\dfrac{9}{20}-\dfrac{1}{10}\right)}{20\left(\dfrac{1}{10}+\dfrac{9}{20}\right)}$

$=\dfrac{20\cdot\dfrac{9}{20}-20\cdot\dfrac{1}{10}}{20\cdot\dfrac{1}{10}+20\cdot\dfrac{9}{20}}$

$=\dfrac{9-2}{2+9}$

$=\dfrac{7}{11}$

33. $\dfrac{1+\dfrac{2}{3}}{1-\dfrac{2}{3}}=\dfrac{3\left(1+\dfrac{2}{3}\right)}{3\left(1-\dfrac{2}{3}\right)}$

$=\dfrac{3\cdot1+3\cdot\dfrac{2}{3}}{3\cdot1-3\cdot\dfrac{2}{3}}$

$=\dfrac{3+2}{3-2}$

$=5$

35. $\dfrac{2+\dfrac{5}{6}}{5-\dfrac{1}{3}}=\dfrac{6\left(2+\dfrac{5}{6}\right)}{6\left(5-\dfrac{1}{3}\right)}$

$=\dfrac{6\cdot2+6\cdot\dfrac{5}{6}}{6\cdot5-6\cdot\dfrac{1}{3}}$

$=\dfrac{12+5}{30-2}$

$=\dfrac{17}{28}$

37. $\dfrac{3+\dfrac{5}{6}}{1+\dfrac{5}{3}}=\dfrac{6\left(3+\dfrac{5}{6}\right)}{6\left(1+\dfrac{5}{3}\right)}$

$=\dfrac{6\cdot3+6\cdot\dfrac{5}{6}}{6\cdot1+6\cdot\dfrac{5}{3}}$

$=\dfrac{18+5}{6+10}$

$=\dfrac{23}{16}$

$=1\dfrac{7}{16}$

39. $\dfrac{\dfrac{1}{3}+\dfrac{3}{4}}{2-\dfrac{1}{6}}=\dfrac{12\left(\dfrac{1}{3}+\dfrac{3}{4}\right)}{12\left(2-\dfrac{1}{6}\right)}$

$=\dfrac{12\cdot\dfrac{1}{3}+12\cdot\dfrac{3}{4}}{12\cdot2-12\cdot\dfrac{1}{6}}$

$=\dfrac{4+9}{24-2}$

$=\dfrac{13}{22}$

41. $\dfrac{\dfrac{5}{6}}{3+\dfrac{2}{3}}=\dfrac{6\left(\dfrac{5}{6}\right)}{6\left(3+\dfrac{2}{3}\right)}$

$=\dfrac{6\cdot\dfrac{5}{6}}{6\cdot3+6\cdot\dfrac{2}{3}}$

$=\dfrac{5}{18+4}$

$=\dfrac{5}{22}$

43.

$$\dfrac{2\frac{1}{2}+\frac{1}{2}}{3\frac{3}{5}-\frac{2}{5}}=\dfrac{\frac{5}{2}+\frac{1}{2}}{\frac{18}{5}-\frac{2}{5}}$$

$$=\dfrac{\frac{6}{2}}{\frac{16}{5}}=\dfrac{\frac{3}{1}}{\frac{16}{5}}$$

$$=3\div\frac{16}{5}$$

$$=3\cdot\frac{5}{16}$$

$$=\frac{15}{16}$$

45.

$$\dfrac{2+1\frac{2}{3}}{3\frac{5}{6}-1}=\dfrac{2+\frac{5}{3}}{\frac{23}{6}-1}$$

$$=\dfrac{6\left(2+\frac{5}{3}\right)}{6\left(\frac{23}{6}-1\right)}$$

$$=\dfrac{12+10}{23-6}$$

$$=\frac{22}{17}$$

$$=1\frac{5}{17}$$

47.

$$\dfrac{3\frac{1}{4}-2\frac{1}{2}}{5\frac{3}{4}+1\frac{1}{2}}=\dfrac{\frac{13}{4}-\frac{5}{2}}{\frac{23}{4}+\frac{3}{2}}$$

$$=\dfrac{4\left(\frac{13}{4}-\frac{5}{2}\right)}{4\left(\frac{23}{4}+\frac{3}{2}\right)}$$

$$=\dfrac{13-10}{23+6}$$

$$=\frac{3}{29}$$

49.

$$\dfrac{3\frac{1}{4}+5\frac{1}{6}}{2\frac{1}{3}+3\frac{1}{4}}=\dfrac{\frac{13}{4}+\frac{31}{6}}{\frac{7}{3}+\frac{13}{4}}$$

$$=\dfrac{12\left(\frac{13}{4}+\frac{31}{6}\right)}{12\left(\frac{7}{3}+\frac{13}{4}\right)}$$

$$=\dfrac{39+62}{28+39}$$

$$=\frac{101}{67}$$

$$=1\frac{34}{67}$$

51.

$$\dfrac{6\frac{2}{3}+7\frac{3}{4}}{8\frac{1}{2}+9\frac{7}{8}}=\dfrac{\frac{20}{3}+\frac{31}{4}}{\frac{17}{2}+\frac{79}{8}}$$

$$=\dfrac{24\left(\frac{20}{3}+\frac{31}{4}\right)}{24\left(\frac{17}{2}+\frac{79}{8}\right)}$$

$$=\dfrac{160+186}{204+237}$$

$$=\frac{346}{441}$$

53. $2\left(2\dfrac{1}{5}+\dfrac{3}{6}\right)=2\left(\dfrac{11}{5}+\dfrac{3}{6}\right)$

$\qquad = 2\left(\dfrac{11\cdot 6}{5\cdot 6}+\dfrac{3\cdot 5}{6\cdot 5}\right)$

$\qquad = 2\left(\dfrac{66}{30}+\dfrac{15}{30}\right)$

$\qquad = 2\cdot\dfrac{81}{30}$

$\qquad = \dfrac{81}{15}$

$\qquad = 5\dfrac{6}{15}$

$\qquad = 5\dfrac{2}{5}$

55. $\left(\dfrac{3}{4}+2\right)+5\dfrac{1}{4}=\dfrac{3}{4}+\dfrac{2\cdot 4}{1\cdot 4}+\dfrac{21}{4}$

$\qquad = \dfrac{3+8+21}{4}$

$\qquad = \dfrac{32}{4}$

$\qquad = 8$

Applying the Concepts

57. Usual purchase:

$32\dfrac{1}{2}=\quad 32\dfrac{1\cdot 3}{2\cdot 3}=\quad 32\dfrac{3}{6}$

$+\,25\dfrac{1}{3}=+\,25\dfrac{1\cdot 2}{3\cdot 2}=+\,25\dfrac{2}{6}$

$\overline{\qquad\qquad\qquad\qquad\qquad\quad 57\dfrac{5}{6}}$ yds

Double the order:

$2\left(57\dfrac{5}{6}\right)=2\left(\dfrac{347}{6}\right)=\dfrac{347}{3}=115\dfrac{2}{3}$ yds

59. Mean $=\dfrac{\text{Sum of the gains/losses}}{\text{Number of gains/losses}}$

$\qquad = \dfrac{\dfrac{3}{4}+\dfrac{9}{16}+\dfrac{3}{32}+\dfrac{7}{32}+\dfrac{1}{16}}{5}$

$\qquad = \dfrac{\dfrac{24}{32}+\dfrac{18}{32}+\dfrac{3}{32}+\dfrac{7}{32}+\dfrac{2}{32}}{5}$

$\qquad = \dfrac{\dfrac{54}{32}}{5}$

$\qquad = \dfrac{54}{32}\cdot\dfrac{1}{5}$

$\qquad = \dfrac{27}{16}\cdot\dfrac{1}{5}$

$\qquad = \dfrac{27}{80}$

The mean gain for the week is $\$\dfrac{27}{80}$.

Review Problems

61. $\dfrac{3}{4}\cdot\dfrac{8}{9}=\dfrac{\cancel{3}\cdot\cancel{4}\cdot 2}{\cancel{4}\cdot\cancel{3}\cdot 3}=\dfrac{2}{3}$

63. $\dfrac{2}{3}\div 4=\dfrac{2}{3}\cdot\dfrac{1}{4}=\dfrac{\cancel{2}\cdot 1}{3\cdot\cancel{2}\cdot 2}=\dfrac{1}{6}$

65. $\dfrac{3}{7}-\dfrac{2}{7}=\dfrac{3-2}{7}=\dfrac{1}{7}=\dfrac{1}{7}$

67. $10-\dfrac{2}{9}=\dfrac{10\cdot 9}{1\cdot 9}-\dfrac{2}{9}=\dfrac{90}{9}-\dfrac{2}{9}=\dfrac{88}{9}=9\dfrac{7}{9}$

Chapter 2 Review

1. $\dfrac{6}{8} = \dfrac{\cancel{2} \cdot 3}{\cancel{2} \cdot 4} = \dfrac{3}{4}$

3. $\dfrac{110}{70} = \dfrac{\cancel{10} \cdot 11}{\cancel{10} \cdot 7} = \dfrac{11}{7}$

5. $\dfrac{1}{5}(5) = \dfrac{1}{5} \cdot 5 = \dfrac{\cancel{1}}{\cancel{5}} \cdot \dfrac{\cancel{5}}{\cancel{1}} = 1$

7. $\dfrac{96}{25} \cdot \dfrac{15}{98} \cdot \dfrac{35}{54}$

$= \dfrac{\left(\cancel{2} \cdot \cancel{2} \cdot 2 \cdot 2 \cdot 2 \cdot \cancel{3}\right) \cdot \left(\cancel{3} \cdot \cancel{5}\right) \cdot \left(\cancel{5} \cdot \cancel{7}\right)}{\left(\cancel{5} \cdot \cancel{5}\right)\left(\cancel{2} \cdot \cancel{7} \cdot 7\right)\left(\cancel{2} \cdot \cancel{3} \cdot \cancel{3} \cdot 3\right)}$

$= \dfrac{8}{21}$

9. $\dfrac{8}{9} \div \dfrac{4}{3} = \dfrac{8}{9} \cdot \dfrac{3}{4} = \dfrac{2 \cdot \cancel{4} \cdot \cancel{3}}{3 \cdot \cancel{3} \cdot \cancel{4}} = \dfrac{2}{3}$

11. $\dfrac{15}{36} \div \dfrac{10}{9} = \dfrac{15}{36} \cdot \dfrac{9}{10} = \dfrac{3 \cdot \cancel{5} \cdot \cancel{9}}{4 \cdot \cancel{9} \cdot \cancel{5} \cdot 2} = \dfrac{3}{8}$

13. $\dfrac{6}{8} - \dfrac{2}{8} = \dfrac{6-2}{8} = \dfrac{4}{8} = \dfrac{1}{2}$

15. $3 + \dfrac{1}{2} = \dfrac{3 \cdot 2}{1 \cdot 2} + \dfrac{1}{2}$

$= \dfrac{6}{2} + \dfrac{1}{2}$

$= \dfrac{6+1}{2}$

$= \dfrac{7}{2}$

$= \dfrac{7}{2}$

17. $\left.\begin{array}{l} 126 = 2 \cdot 3 \cdot 3 \cdot 7 \\ 84 = 2 \cdot 2 \cdot 3 \cdot 7 \end{array}\right\} \text{LCD} = 2 \cdot 2 \cdot 3 \cdot 3 \cdot 7 = 252$

$\dfrac{11}{126} - \dfrac{5}{84} = \dfrac{11 \cdot 2}{126 \cdot 2} - \dfrac{5 \cdot 3}{84 \cdot 3}$

$= \dfrac{22}{252} - \dfrac{15}{252}$

$= \dfrac{7}{252}$

$= \dfrac{\cancel{7}}{\cancel{7} \cdot 36}$

$= \dfrac{1}{36}$

19. $3\dfrac{5}{8} = \dfrac{8 \cdot 3 + 5}{8} = \dfrac{29}{8}$

21. $\dfrac{15}{4} : \quad 4\overline{)\begin{array}{l} 3 \\ 15 \\ \underline{12} \\ 3 \end{array}} \quad \text{so} \quad \dfrac{15}{4} = 3 + \dfrac{3}{4} = 3\dfrac{3}{4}$

23. $2 \div 3\dfrac{1}{4} = 2 \div \dfrac{13}{4} = 2 \cdot \dfrac{4}{13} = \dfrac{2 \cdot 4}{1 \cdot 13} = \dfrac{8}{13}$

25. $6 \cdot 2\dfrac{1}{2} \cdot \dfrac{4}{5} = \dfrac{6}{1} \cdot \dfrac{5}{2} \cdot \dfrac{4}{5} = \dfrac{3 \cdot \cancel{2} \cdot \cancel{5} \cdot 4}{1 \cdot \cancel{2} \cdot \cancel{5}} = 12$

27. $\begin{array}{r} 8\dfrac{2}{3} = \\ + 9\dfrac{1}{4} = \\ \hline \end{array} \begin{array}{r} 8\dfrac{2 \cdot 4}{3 \cdot 4} = \\ + 9\dfrac{1 \cdot 3}{4 \cdot 3} = \\ \hline \end{array} \begin{array}{r} 8\dfrac{8}{12} \\ + 9\dfrac{3}{12} \\ \hline 17\dfrac{11}{12} \end{array}$

29. $3 + 2\left(4\frac{1}{3}\right) = 3 + 2\left(\frac{13}{3}\right)$

$$= 3 + \frac{26}{3}$$

$$= \frac{3 \cdot 3}{1 \cdot 3} + \frac{26}{3}$$

$$= \frac{9}{3} + \frac{26}{3}$$

$$= \frac{35}{3}$$

$$= 11\frac{2}{3}$$

31. $\dfrac{1 + \dfrac{2}{3}}{1 - \dfrac{2}{3}} = \dfrac{3\left(1 + \dfrac{2}{3}\right)}{3\left(1 - \dfrac{2}{3}\right)} = \dfrac{3 + 2}{3 - 2} = \dfrac{5}{1} = 5$

33. $\dfrac{\dfrac{7}{8} - \dfrac{1}{2}}{\dfrac{1}{4} + \dfrac{1}{2}} = \dfrac{8\left(\dfrac{7}{8} - \dfrac{1}{2}\right)}{8\left(\dfrac{1}{4} + \dfrac{1}{2}\right)} = \dfrac{7 - 4}{2 + 4} = \dfrac{3}{6} = \dfrac{1}{2}$

35. $\dfrac{1}{10}$ of $200 =$ number of defective items

$$\frac{1}{10} \cdot 200 = \frac{1 \cdot 200}{10 \cdot 1} = \frac{1 \cdot \cancel{10} \cdot 20}{\cancel{10} \cdot 1} = \frac{20}{1} = 20$$

There are 20 defective items.

37. $3\left(2\frac{1}{4} + \frac{3}{4}\right) = 3\left(2 + \frac{1}{4} + \frac{3}{4}\right)$

$$= 3\left(2 + \frac{4}{4}\right)$$

$$= 3(2 + 1)$$

$$= 3(3)$$

$$= 9$$

39. Find the fraction of the recipe to be made:

$$\frac{\text{Cookies needed}}{\text{Total cookies}} = \frac{36}{48} = \frac{3}{4} \text{ of recipe}$$

$\dfrac{3}{4}$ of $2\dfrac{1}{2}$ cups $=$ flour needed

$$\frac{3}{4} \cdot 2\frac{1}{2} = \frac{3}{4} \cdot \frac{5}{2} = \frac{3 \cdot 5}{4 \cdot 2} = \frac{15}{8} = 1\frac{7}{8} \text{ cups}$$

41. $3 \cdot 3\dfrac{1}{2} = 3 \cdot \dfrac{7}{2} = \dfrac{3 \cdot 7}{1 \cdot 2} = \dfrac{21}{2} = 10\dfrac{1}{2}$

$10\dfrac{1}{2}$ tbsp will be used.

43. $A = \dfrac{1}{2}bh$ $\qquad\qquad P = 10\dfrac{2}{5}$

$$= \frac{1}{2} \cdot 12\frac{3}{5} \cdot 4 \qquad\qquad 12\frac{3}{5}$$

$$= \frac{1}{2} \cdot \frac{63}{5} \cdot \frac{4}{1} \qquad\qquad \frac{+\ 5}{}$$

$$= \frac{1 \cdot 63 \cdot \cancel{2} \cdot 2}{\cancel{2} \cdot 5 \cdot 1} \qquad\quad 27\frac{5}{5} = 28 \text{ ft}$$

$$= \frac{126}{5}$$

$$= 25\frac{1}{5} \text{ ft}^2$$

Chapters 1-2 Cumulative Review

1. $(6+2)+(3+6) = 8+9 = 17$

3.
$$\begin{array}{r} 1,985 \\ -\ 141 \\ \hline 1,844 \end{array}$$

5. $\dfrac{5}{8} - \dfrac{3}{8} = \dfrac{5-3}{8} = \dfrac{2}{8} = \dfrac{1}{4}$

7. $\dfrac{1}{2} + \dfrac{1}{4} = \dfrac{1 \cdot 2}{2 \cdot 2} + \dfrac{1}{4} = \dfrac{2}{4} + \dfrac{1}{4} = \dfrac{3}{4}$

9.
$$\begin{array}{r} 5,280 \\ \times\ 26 \\ \hline 31,680 \\ 105,600 \\ \hline 137,280 \end{array}$$

11. $3 \cdot 10^2 + 5 \cdot 10 + 4 = 3 \cdot 100 + 5 \cdot 10 + 4$
$$= 300 + 50 + 4$$
$$= 354$$

13. $\dfrac{1}{2} \cdot \dfrac{4}{5} = \dfrac{1 \cdot \cancel{2} \cdot 2}{\cancel{2} \cdot 5} = \dfrac{2}{5}$

15. $3^2 + 4^2 = 9 + 16 = 25$

17. $3 \div \dfrac{1}{2} = \dfrac{3}{1} \cdot \dfrac{2}{1} = 6$

19. $\dfrac{11}{77} = \dfrac{\cancel{11}}{\cancel{11} \cdot 7} = \dfrac{1}{7}$

21. $121 \div 11 \div 11 = (121 \div 11) \div 11 = 11 \div 11 = 1$

23. $\left(\dfrac{3}{5} + \dfrac{2}{25}\right) - \dfrac{4}{125} = \dfrac{3 \cdot 25}{5 \cdot 25} + \dfrac{2 \cdot 5}{25 \cdot 5} - \dfrac{4}{125}$
$$= \dfrac{75}{125} + \dfrac{10}{125} - \dfrac{4}{125}$$
$$= \dfrac{81}{125}$$

25. $2\dfrac{1}{3}\left(1\dfrac{1}{4} \div \dfrac{1}{5}\right) = \dfrac{7}{3} \cdot \dfrac{5}{4} \div \dfrac{1}{5}$
$$= \dfrac{7}{3} \cdot \dfrac{5}{4} \cdot \dfrac{5}{1}$$
$$= \dfrac{175}{12}$$
$$= 14\dfrac{7}{12}$$

27. $\dfrac{2}{3} + \dfrac{1}{9} + \dfrac{3}{4} = \dfrac{2 \cdot 12}{3 \cdot 12} + \dfrac{1 \cdot 4}{9 \cdot 4} + \dfrac{3 \cdot 9}{4 \cdot 9}$
$$= \dfrac{24}{36} + \dfrac{4}{36} + \dfrac{27}{36}$$
$$= \dfrac{55}{36}$$

29. $\dfrac{7}{9}$ of $3\dfrac{1}{4} = \dfrac{7}{9} \cdot 3\dfrac{1}{4} = \dfrac{7}{9} \cdot \dfrac{13}{4} = \dfrac{91}{36} = 2\dfrac{19}{36}$

31.

Separate into 2 regions.
Area I $= lw = 10 \cdot 6 = 60$ in^2
Area II $= lw = 12 \cdot 11 = 132$ in^2
Total Area = Area I + Area II
$$= 60 \text{ in}^2 + 132 \text{ in}^2 = 192 \text{ in}^2$$

33. $\dfrac{111}{19}$: $19\overline{)111}$ so $\dfrac{111}{19} = 5 + \dfrac{16}{19} = 5\dfrac{16}{19}$

$\dfrac{\;5\;}{}$

$\underline{95}$

16

35. $\dfrac{14}{49} = \dfrac{\cancel{7} \cdot 2}{\cancel{7} \cdot 7} = \dfrac{2}{7}$

37. $\dfrac{1}{2} \cdot \dfrac{4}{9} + \dfrac{4}{5} = \dfrac{1 \cdot \cancel{2} \cdot 2}{\cancel{2} \cdot 9} + \dfrac{4}{5}$

$ = \dfrac{2}{9} + \dfrac{4}{5}$

$ = \dfrac{2 \cdot 5}{9 \cdot 5} + \dfrac{4 \cdot 9}{5 \cdot 9}$

$ = \dfrac{10}{45} + \dfrac{36}{45}$

$ = \dfrac{46}{45} \text{ or } 1\dfrac{1}{45}$

39. $\dfrac{5}{6}$ of 36 exposures + $\dfrac{5}{6}$ of 2 rolls of 24

exposures = pictures taken

$\dfrac{5}{6} \cdot 36 + \dfrac{5}{6} \cdot 2 \cdot 24 = \dfrac{5}{6} \cdot \dfrac{36}{1} + \dfrac{5}{6} \cdot \dfrac{48}{1}$

$ = 30 + 40$

$ = 70$

She took 70 pictures.

Chapter 2 Test

1. Answers will vary.

$\dfrac{1}{8}$: 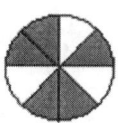 $\dfrac{3}{8}$: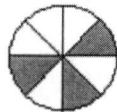

$\dfrac{5}{8}$: $\dfrac{7}{8}$:

3. $\dfrac{3}{5}(30) = \dfrac{3}{5} \cdot \dfrac{30}{1} = \dfrac{3 \cdot \cancel{5} \cdot 6}{\cancel{5} \cdot 1} = 18$

5. $\dfrac{5}{18} \div \dfrac{15}{16} = \dfrac{5}{18} \cdot \dfrac{16}{15} = \dfrac{\cancel{5} \cdot \cancel{2} \cdot 8}{\cancel{2} \cdot 9 \cdot \cancel{5} \cdot 3} = \dfrac{8}{27}$

7. $\dfrac{3}{10} + \dfrac{1}{10} = \dfrac{3+1}{10} = \dfrac{4}{10} = \dfrac{2}{5}$

9. $4 + \dfrac{3}{5} = \dfrac{4 \cdot 5}{1 \cdot 5} + \dfrac{3}{5}$

$\qquad = \dfrac{20}{5} + \dfrac{3}{5}$

$\qquad = \dfrac{20+3}{5}$

$\qquad = \dfrac{23}{5}$

11. $\left.\begin{array}{l} 6 = 2 \cdot 3 \\ 9 = 3 \cdot 3 \\ 4 = 2 \cdot 2 \end{array}\right\}$ LCD $= 2 \cdot 2 \cdot 3 \cdot 3 = 36$

$\dfrac{5}{6} + \dfrac{2}{9} + \dfrac{1}{4} = \dfrac{5 \cdot 6}{6 \cdot 6} + \dfrac{2 \cdot 4}{9 \cdot 4} + \dfrac{1 \cdot 9}{4 \cdot 9}$

$\qquad = \dfrac{30}{36} + \dfrac{8}{36} + \dfrac{9}{36}$

$\qquad = \dfrac{47}{36}$ or $1\dfrac{11}{36}$

13. $5\overline{)43}$ so $\dfrac{43}{5} = 8 + \dfrac{3}{5} = 8\dfrac{3}{5}$
$\quad\ \underline{40}$
$\quad\ \ 3$

with the long division showing 8 above, 40 below 43, remainder 3.

15. $6 \div 1\dfrac{1}{3} = 6 \div \dfrac{4}{3} = \dfrac{6}{1} \cdot \dfrac{3}{4} = \dfrac{\cancel{2} \cdot 3 \cdot 3}{1 \cdot \cancel{2} \cdot 2} = \dfrac{9}{2} = 4\dfrac{1}{2}$

17. $\begin{array}{llll} 5\dfrac{1}{6} = & 5\dfrac{1}{6} = & 5\dfrac{1}{6} = & 4\dfrac{7}{6} \\ -1\dfrac{1}{2} = & -1\dfrac{1 \cdot 3}{2 \cdot 3} = & -1\dfrac{3}{6} = & -1\dfrac{3}{6} \end{array}$

$\qquad\qquad\qquad\qquad\qquad 3\dfrac{4}{6} = 3\dfrac{2}{3}$

19. $\left(2\dfrac{1}{3} + \dfrac{1}{2}\right)\left(3\dfrac{2}{3} - \dfrac{1}{6}\right) = \left(\dfrac{7}{3} + \dfrac{1}{2}\right)\left(\dfrac{11}{3} - \dfrac{1}{6}\right)$

$\qquad = \left(\dfrac{14}{6} + \dfrac{3}{6}\right)\left(\dfrac{22}{6} - \dfrac{1}{6}\right)$

$\qquad = \left(\dfrac{17}{6}\right)\left(\dfrac{21}{6}\right)$

$\qquad = \dfrac{17 \cdot \cancel{3} \cdot 7}{2 \cdot \cancel{3} \cdot 2 \cdot 3}$

$\qquad = \dfrac{119}{12}$

$\qquad = 9\dfrac{11}{12}$

21. $\dfrac{1}{3}$ of 120 = number spoiled

$\dfrac{1}{3} \cdot 120 = \dfrac{1 \cdot \cancel{3} \cdot 40}{\cancel{3} \cdot 1} = \dfrac{40}{1} = 40$

40 grapefruit are spoiled.

23. $2\left(4\dfrac{2}{3}\right) = \left(\dfrac{2}{1}\right)\left(\dfrac{14}{3}\right) = \dfrac{28}{3} = 9\dfrac{1}{3}$

$9\dfrac{1}{3}$ cups will be used.

25.

$$A = \frac{1}{2}bh \qquad\qquad P = 8\frac{1}{3}$$

$$= \frac{1}{2} \cdot 9\frac{1}{3} \cdot 5 \qquad\qquad 5\frac{2}{3}$$

$$= \frac{1}{2} \cdot \frac{28}{3} \cdot \frac{5}{1} \qquad\qquad +\ 9\frac{1}{3}$$

$$= \frac{1 \cdot \cancel{2} \cdot 14 \cdot 5}{\cancel{2} \cdot 3 \cdot 1} \qquad 22\frac{4}{3} = 23\frac{1}{3} \text{ ft}$$

$$= \frac{70}{3} = 23\frac{1}{3} \text{ ft}^2$$

Chapter 2: A Glimpse of Algebra

1. $\dfrac{x}{4} + \dfrac{3}{4} = \dfrac{x+3}{4}$

3. $\dfrac{x+6}{5} - \dfrac{4}{5} = \dfrac{x+6-4}{5} = \dfrac{x+2}{5}$

5. $\dfrac{5x}{8} + \dfrac{2x}{8} = \dfrac{5x+2x}{8} = \dfrac{7x}{8}$

7. $\dfrac{6}{x} - \dfrac{4}{x} = \dfrac{6-4}{x} = \dfrac{2}{x}$

9. $\text{LCD} = 2 \cdot 3 = 6$

$\dfrac{x}{2} + \dfrac{1}{3} = \dfrac{x \cdot 3}{2 \cdot 3} + \dfrac{1 \cdot 2}{3 \cdot 2}$

$= \dfrac{3x}{6} + \dfrac{2}{6}$

$= \dfrac{3x+2}{6}$

11. $\left.\begin{array}{l} 2 = 2 \\ 4 = 2 \cdot 2 \end{array}\right\} \text{LCD} = 2 \cdot 2 = 4$

$\dfrac{x}{2} - \dfrac{1}{4} = \dfrac{x \cdot 2}{2 \cdot 2} - \dfrac{1}{4}$

$= \dfrac{2x}{4} - \dfrac{1}{4}$

$= \dfrac{2x-1}{4}$

13. $\text{LCD} = x \cdot 4 = 4x$

$\dfrac{3}{x} + \dfrac{3}{4} = \dfrac{3 \cdot 4}{x \cdot 4} + \dfrac{3 \cdot x}{4 \cdot x}$

$= \dfrac{12}{4x} + \dfrac{3x}{4x}$

$= \dfrac{12+3x}{4x}$

15. $\text{LCD} = 5 \cdot x = 5x$

$\dfrac{4}{5} - \dfrac{1}{x} = \dfrac{4 \cdot x}{5 \cdot x} - \dfrac{1 \cdot 5}{x \cdot 5}$

$= \dfrac{4x}{5x} - \dfrac{5}{5x}$

$= \dfrac{4x-5}{5x}$

17. $\text{LCD} = 3 \cdot x \cdot 4 = 12x$

$\dfrac{1}{3} + \dfrac{2}{x} + \dfrac{1}{4} = \dfrac{1 \cdot 4x}{3 \cdot 4x} + \dfrac{2 \cdot 12}{x \cdot 12} + \dfrac{1 \cdot 3x}{4 \cdot 3x}$

$= \dfrac{4x}{12x} + \dfrac{24}{12x} + \dfrac{3x}{12x}$

$= \dfrac{7x+24}{12x}$

19. $\left.\begin{array}{l} 2 = 2 \\ 4 = 2 \cdot 2 \\ x = x \end{array}\right\} \text{LCD} = 2 \cdot 2 \cdot x = 4x$

$\dfrac{1}{2} + \dfrac{1}{x} + \dfrac{1}{4} = \dfrac{1 \cdot 2x}{2 \cdot 2x} + \dfrac{1 \cdot 4}{x \cdot 4} + \dfrac{1 \cdot x}{4 \cdot x}$

$= \dfrac{2x}{4x} + \dfrac{4}{4x} + \dfrac{x}{4x}$

$= \dfrac{3x+4}{4x}$

21. $3 + \dfrac{x}{4} = \dfrac{3 \cdot 4}{1 \cdot 4} + \dfrac{x}{4} = \dfrac{12}{4} + \dfrac{x}{4} = \dfrac{12+x}{4}$

23. $5 - \dfrac{x}{2} = \dfrac{5 \cdot 2}{1 \cdot 2} - \dfrac{x}{2} = \dfrac{10}{2} - \dfrac{x}{2} = \dfrac{10-x}{2}$

25. $4 - \dfrac{a}{7} = \dfrac{4 \cdot 7}{1 \cdot 7} - \dfrac{a}{7} = \dfrac{28}{7} - \dfrac{a}{7} = \dfrac{28-a}{7}$

27. $1 + \dfrac{2a}{5} = \dfrac{1 \cdot 5}{1 \cdot 5} + \dfrac{2a}{5} = \dfrac{5}{5} + \dfrac{2a}{5} = \dfrac{5+2a}{5}$

29. $8 + \dfrac{3}{x} = \dfrac{8 \cdot x}{1 \cdot x} + \dfrac{3}{x} = \dfrac{8x}{x} + \dfrac{3}{x} = \dfrac{8x+3}{x}$

31. $2 - \dfrac{5}{x} = \dfrac{2 \cdot x}{1 \cdot x} - \dfrac{5}{x} = \dfrac{2x}{x} - \dfrac{5}{x} = \dfrac{2x-5}{x}$

33. $x + \dfrac{3}{4} = \dfrac{x \cdot 4}{1 \cdot 4} + \dfrac{3}{4} = \dfrac{4x}{4} + \dfrac{3}{4} = \dfrac{4x+3}{4}$

35. $x + \dfrac{2x}{9} = \dfrac{x \cdot 9}{1 \cdot 9} + \dfrac{2x}{9} = \dfrac{9x}{9} + \dfrac{2x}{9} = \dfrac{11x}{9}$

37. $a - \dfrac{4a}{7} = \dfrac{a \cdot 7}{1 \cdot 7} - \dfrac{4a}{7} = \dfrac{7a}{7} - \dfrac{4a}{7} = \dfrac{3a}{7}$

39. $2x - \dfrac{3x}{4} = \dfrac{2x \cdot 4}{1 \cdot 4} - \dfrac{3x}{4} = \dfrac{8x}{4} - \dfrac{3x}{4} = \dfrac{5x}{4}$

Chapter 3 Preview

1.
$$\begin{array}{r} 25{,}430 \\ 2{,}897 \\ +379{,}600 \\ \hline 407{,}927 \end{array}$$

3.
$$\begin{array}{r} 2{,}000 \\ -1{,}564 \\ \hline 436 \end{array}$$

5.
$$\begin{array}{r} 305 \\ \times 436 \\ \hline 1830 \\ 9150 \\ \underline{122000} \\ 132{,}980 \end{array}$$

7. $480 + 12(32)^2 = 480 + 12(1{,}024)$
$$= 480 + 12{,}288$$
$$= 12{,}768$$

9.
$$\begin{array}{r} 1{,}848 \\ 27\overline{)49{,}896} \\ \underline{27\downarrow} \\ 228 \\ \underline{216\downarrow} \\ 129 \\ \underline{108\downarrow} \\ 216 \\ \underline{216} \\ 0 \end{array}$$

11. $5^2 + 7^2 = 25 + 49 = 74$

13. $\dfrac{3}{4}\left(\dfrac{2}{5}\right) = \dfrac{3 \cdot \cancel{2}}{\cancel{2} \cdot 2 \cdot 5} = \dfrac{3}{10}$

15. $\dfrac{38}{100} = \dfrac{\cancel{2} \cdot 19}{\cancel{2} \cdot 50} = \dfrac{19}{50}$

17. $\dfrac{1}{2}\left(2\dfrac{5}{8}\right)\left(1\dfrac{1}{4}\right) = \dfrac{1}{2} \cdot \dfrac{21}{8} \cdot \dfrac{5}{4} = \dfrac{105}{64} = 1\dfrac{41}{64}$

19. $48 = 2 \cdot 2 \cdot 2 \cdot 2 \cdot 3 = 2^4 \cdot 3$

Chapter 3 Pretest

1. 4.013 in words:
   four and thirteen thousandths

3. Thirty-four hundredths is 0.34

5. Add trailing 0's so all numbers have an
   equal number of digits. Then order:
   0.04
   0.40
   0.41
   0.45
   0.50
   0.51

7. 7.36
   $\underline{+\ 8.05}$
   15.41

9. 3.6
   $\underline{\times\ 2.7}$
   252
   $\underline{720}$
   9.72

11. $\begin{array}{r} 4.12 \\ 32\overline{)131.84} \\ \underline{128\!\downarrow} \\ 38 \\ \underline{32\!\downarrow} \\ 64 \\ \underline{64} \\ 0 \end{array}$

13. $\begin{array}{r} 0.42 \\ 50\overline{)21.00} \\ \underline{200\!\downarrow} \\ 100 \\ \underline{100} \\ 0 \end{array}$

15. $\$20.00$
    $\underline{-16.93}$
    $\$3.07$

17. $c = \sqrt{a^2 + b^2}$
    $= \sqrt{9^2 + 12^2}$
    $= \sqrt{81 + 144}$
    $= \sqrt{225}$
    $= 15$

Section 3.1

1. 0.3 in words:
    three tenths

3. 0.015 in words:
    fifteen thousandths

5. 3.4 in words:
    three and four tenths

7. 52.7 in words:
    fifty-two and seven tenths

9. $405.36 = 405\dfrac{36}{100}$

11. $9.009 = 9\dfrac{9}{1,000}$

13. $1.234 = 1\dfrac{234}{1,000}$

15. $0.00305 = \dfrac{305}{100,000}$

17. 458.327

    tens

19. 29.52

    tenths

21. 0.00375
    ↑
    hundred thousandths

23. 275.01

    ones

25. 539.76
    ↑
    hundreds

27. Fifty-five hundredths is 0.55

29. Six and nine tenths is 6.9

31. Eleven and eleven hundredths is 11.11

33. One hundred and two hundredths is 100.02

35. Three thousand and three thousandths is
    3,000.003

37. Look at the digit to the right of the one you
    are rounding to:
    Whole number   47.**5**479
    $5 \geq 5$ so add 1 to the ones place and delete
    all digits to the right   48

    Tenth   47.5**4**79
    $4 < 5$ so leave the tenths place as is and
    delete all digits to the right   47.5

    Hundredth   47.54**7**9
    $7 \geq 5$ so add 1 to the hundredths place and
    delete all digits to the right   47.55

    Thousandth   47.547**9**
    $9 \geq 5$ so add 1 to the thousandths place and
    delete all digits to the right   47.548

39. Look at the digit to the right of the one you are rounding to:
Whole number 0.**8**175
8 ≥ 5 so add 1 to the ones place and delete all digits to the right   1

Tenth   0.8**1**75
1 < 5 so leave the tenths place as is and delete all digits to the right   0.8

Hundredth   0.81**7**5
7 ≥ 5 so add 1 to the hundredths place and delete all digits to the right   0.82

Thousandth   0.817**5**
5 ≥ 5 so add 1 to the thousandths place and delete all digits to the right   0.818

41. Look at the digit to the right of the one you are rounding to:
Whole number 0.**1**562
1 < 5 so leave the ones place as is and delete all digits to the right   0

Tenth   0.1**5**62
5 ≥ 5 so add 1 to the tenths place and delete all digits to the right   0.2

Hundredth   0.15**6**2
6 ≥ 5 so add 1 to the hundredths place and delete all digits to the right   0.16

Thousandth   0.156**2**
2 < 5 so leave the thousandths place as is and delete all digits to the right   0.156

43. Look at the digit to the right of the one you are rounding to:
Whole number 2,789.**3**241
3 < 5 so leave the ones place as is and delete all digits to the right   2,789

Tenth   2,789.3**2**41
2 < 5 so leave the tenths place as is and delete all digits to the right   2,789.3

Hundredth   2,789.32**4**1
4 < 5 so leave the hundredths place as is and delete all digits to the right   2,789.32

Thousandth   2,789.324**1**
1 < 5 so leave the thousandths place as is and delete all digits to the right   2,789.324

45. Look at the digit to the right of the one you are rounding to:
Whole number 99.**9**999
9 ≥ 5 so add 1 to the ones place (and carry) and delete all digits to the right   100

Tenth   99.9**9**99
9 ≥ 5 so add 1 to the tenths place (and carry) and delete all digits to the right
                                                    100.0
Hundredth   99.99**9**9
9 ≥ 5 so add 1 to the hundredths place (and carry) and delete all digits to the right
                                                    100.00
Thousandth   99.999**9**
9 ≥ 5 so add 1 to the thousandths place (and carry) and delete all digits to the right
                                                    100.000

Applying the Concepts

47. 3.11 in words:
Three and eleven hundredths
2.5 in words:
Two and five tenths

49. Look at the digit to the right of the one you are rounding to:
Hundredth   186,282.39$\underline{7}$6
$7 \geq 5$ so add 1 to the hundredths place (and carry) and delete all digits to the right
                    186,282.40

51. 0.15 in words:
Fifteen hundredths

53. For each date, read the corresponding price at the top of the bar as measured by the vertical axis.

Date	Price (dollars)
4/5/2004	2.126
4/12/04	2.157
4/19/04	2.148
4/26/04	2.124

55. To compare the numbers, add trailing zeros where needed so all numbers have the same amount of digits. Then arrange from smallest to largest by reading from left to right:

Original Set	Ordered Set
0.020	0.002
0.050	0.005
0.025	0.020
0.052	0.025
0.005	0.050
0.002	0.052

57. Look at the digit to the right of the tens place:
7.4$\underline{5}$1 $5 \geq 5$ so add 1 to the tens place and delete all digits to the right   7.5
7.4$\underline{4}$9 $4 < 5$ so leave the tens place as is and delete all digits to the right   7.4
7.5$\underline{4}$ $4 < 5$ so leave the tens place as is and delete all digits to the right   7.5
7.5$\underline{6}$ $6 \geq 5$ so add 1 to the tens place and delete all digits to the right   7.6

The answers are 7.451 and 7.54

59. $0.25 = \dfrac{25}{100} = \dfrac{1}{4}$

61. $0.35 = \dfrac{35}{100} = \dfrac{7}{20}$

63. $0.125 = \dfrac{125}{1,000} = \dfrac{1}{8}$

65. $0.625 = \dfrac{625}{1,000} = \dfrac{5}{8}$

67. $0.875 = \dfrac{875}{1,000} = \dfrac{7}{8}$

Estimating

69. Add trailing zeros so that both numbers have an equal number of digits.
9.90
9.99
By reading the digits from left to right, the larger number is 9.99, which is closer to 10.

71. Add trailing zeros so that both numbers have an equal number of digits.
10.50
10.05
By reading the digits from left to right, the smaller number is 10.05, which is closer to 10.

73. Add trailing zeros so that both numbers have an equal number of digits.
0.50
0.05
By reading the digits from left to right, the smaller number is 0.05, which is closer to 0.

75. Add trailing zeros so that both numbers have an equal number of digits.
0.01
0.02
By reading the digits from left to right, the smaller number is 0.01, which is closer to 0.

Review Problems

77.
$$4\frac{3}{10} = 4\frac{3\cdot 10}{10\cdot 10} = 4\frac{30}{100}$$
$$+\ 2\frac{1}{100} = +\ 2\frac{1}{100} = +\ 2\frac{1}{100}$$
$$6\frac{31}{100}$$

79.
$$8\frac{5}{10} = 8\frac{5\cdot 10}{10\cdot 10} = 8\frac{50}{100}$$
$$-\ 2\frac{4}{100} = -\ 2\frac{4}{100} = -\ 2\frac{4}{100}$$
$$6\frac{46}{100} = 6\frac{23}{50}$$

81.
$$5\frac{1}{10} = 5\frac{1\cdot 100}{10\cdot 100} = 5\frac{100}{1000}$$
$$6\frac{2}{100} = 6\frac{2\cdot 10}{100\cdot 10} = 6\frac{20}{1000}$$
$$+\ 7\frac{3}{1000} = +\ 7\frac{3}{1000} = +\ 7\frac{3}{1000}$$
$$18\frac{123}{1,000}$$

Improving Your Quantitative Literacy

83. a. Reading the graph from left to right, find where the costs are falling:
1990, 1991, 2000, 2002, 2003

b. Reading the graph from left to right, find where the line is the steepest over two years: 1998 to 2000

c. Reading the price graph from left to right, find the steepest drop over a 1 year period: 1985

d. Answers will vary.

Section 3.2

1.   2.91
    + 3.28
    6.19

3.   0.04
    0.31
    + 0.78
    1.13

5.   3.89
    + 2.40
    6.29

7.   Note: 1.81 = 1.810
          2.7 = 2.700
       4.532
       1.810
     + 2.700
       9.042

9.   Note: 5 = 5.000
          2.94 = 2.940
       0.081
       5.000
     + 2.940
       8.021

11. Note: 6.78 = 6.7800
          0.004 = 0.0040
       5.0003
       6.7800
     + 0.0040
      11.7843

13.   7.123
       8.120
     + 9.100
      24.343

15.   9.001
       8.010
     + 7.100
      24.111

17.   89.7854
        3.4000
       65.3500
     + 100.0060
      258.5414

19.   543.21
     + 123.45
      666.66

21.   99.34
      − 88.23
       11.11

23.   5.97
      − 2.40
       3.57

25.   6.30
      − 2.08
       4.22

27.   149.37
      − 28.96
      120.41

29.   45.000
      − 0.067
      44.933

31.

$$8.000$$
$$\underline{-\ 0.327}$$
$$7.673$$

33.

$$765.432$$
$$\underline{-\ 234.567}$$
$$530.865$$

35.

$$34.07$$
$$\underline{-\ 6.18}$$
$$27.89$$

37.

$$40.04$$
$$\underline{-\ 4.40}$$
$$35.64$$

39.

$$768.436$$
$$\underline{-\ 356.998}$$
$$411.438$$

41. $(7.8 - 4.3) + 2.5$

$$= 3.5 + 2.5$$
$$= 6$$

43. $7.8 - (4.3 + 2.5)$

$$= 7.8 - 6.8$$
$$= 1$$

45. $(9.7 - 5.2) - 1.4$

$$= 4.5 - 1.4$$
$$= 3.1$$

47. $9.7 - (5.2 - 1.4)$

$$= 9.7 - 3.8$$
$$= 5.9$$

49. $(8.2 + 0.072) - 5$

$$= 8.272 - 5$$
$$= 3.272$$

51. Let $x =$ the number

$$0.035 + x = 4.036$$
$$x = 4.036 - 0.035$$
$$x = 4.001$$

Applying the Concepts

53. Find the total of the amounts spent at each store:

$$\$25.37$$
$$39.41$$
$$\underline{+\ 52.04}$$
$$\$116.82$$

55. Find the sum of the highest prices for each album:

$$\$20.79$$
$$13.99$$
$$27.28$$
$$13.95$$
$$\underline{+\ 18.81}$$
$$\$94.82$$

57. Total deductions:

$$\$311.93$$
$$158.21$$
$$\underline{+\ 64.72}$$
$$\$534.86$$

Gross Pay: $\$2,105.96$
$\underline{-\text{Total Deductions:}\quad -534.86}$
Take-home Pay $\$1,571.10$

59. $P = 2l + 2w$

$$= 2(1.41 \text{ in}) + 2(0.84 \text{ in})$$
$$= 2.82 \text{ in} + 1.68 \text{ in}$$
$$= 4.5 \text{ in}$$

61. $10.00
$$\begin{array}{r} \$10.00 \\ -\ 4.57 \\ \hline \$5.43 \end{array}$$
They receive $5.43 in change.

63. a.
$$\begin{array}{r} \text{2002 subscribers:}\quad 66.9 \text{ million} \\ -\ \text{1990 subscribers:}\quad -50.5 \text{ million} \\ \hline \text{Difference:}\quad 16.4 \text{ million} \end{array}$$

b.
$$\begin{array}{r} \text{1990 subscribers:}\quad 62.3 \text{ million} \\ -\ \text{2002 subscribers:}\quad -55.2 \text{ million} \\ \hline \text{Difference:}\quad 7.1 \text{ million} \end{array}$$

c. There was a greater increase in cable subscribers with 16.4 million.

65. a. From 2006 on the horizontal axis, draw a vertical line up to the graph. It is slightly more than 15 million.

b. From 2005 on the horizontal axis, draw a vertical line up to the graph. It is slightly less than 10 million.

c. Answers will vary:
$$\begin{array}{r} \text{2006 sales:}\quad 15.5 \\ -\ \text{2005 sales:}\quad -9.5 \\ \hline \text{Increase:}\quad 6.0 \text{ million} \end{array}$$

67. Above each car, draw a bar with height corresponding to the time as measured by the vertical axis.

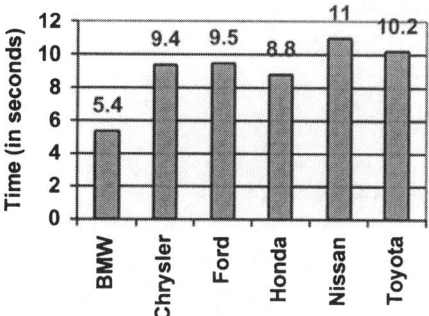

$$\begin{array}{r} \text{Slowest car:}\quad 11.0 \\ -\ \text{Fastest car:}\quad -\ 5.4 \\ \hline \text{Time difference:}\quad 5.6 \text{ seconds} \end{array}$$

Estimating

69.
$$\begin{array}{r} 0.005 \\ +\ 0.090 \\ \hline 0.095 \text{ rounds to } 0.1 \end{array}$$

Review Problems

71. $\dfrac{1}{10} \cdot \dfrac{3}{10} = \dfrac{1 \cdot 3}{10 \cdot 10} = \dfrac{3}{100}$

73. $\dfrac{3}{100} \cdot \dfrac{17}{100} = \dfrac{3 \cdot 17}{100 \cdot 100} = \dfrac{51}{10,000}$

75. $5\left(\dfrac{3}{10}\right) = \dfrac{5}{1} \cdot \dfrac{3}{10} = \dfrac{5 \cdot 3}{1 \cdot 10} = \dfrac{15}{10} = \dfrac{3}{2} = 1\dfrac{1}{2}$

77.
$$\begin{array}{r} 56 \\ \times\ 25 \\ \hline 280 \\ 1,120 \\ \hline 1,400 \end{array}$$

Extending the Concepts

79. $P = 2l + 2w$ let $P = 9.5, l = 2.75$

$9.5 = 2(2.75) + 2w$

$9.5 = 5.5 + 2w$

$9.5 + (-5.5) = 5.5 + 2w + (-5.5)$

$4 = 2w$

$2 = w$

The width is 2 inches.

81.

Amount given cashier:	$20.01
− Total bill:	− 16.76
Change:	$3.25

You receive $3.25 in change, probably as three $1 bills and a quarter.

83. $2.5, 2.75, 3, ...$

The first number in the sequence is 2.5. Each number thereafter comes from adding 0.25 to the previous number, meaning the sequence is *arithmetic*. The next number in the sequence is $3 + 0.25 = 3.25$

Improving Your Quantitative Literacy

85. a.

2006 users:	163.7 million
− 2003 users:	−147.6 million
Increase:	16.1 million

b. The difference between sequential years:

Years	Increase (in millions)
2003 to 2004	$154.2 - 147.5 = 6.7$
2004 to 2005	$159.7 - 154.2 = 5.5$
2005 to 2006	$163.7 - 159.7 = 4.0$
2006 to 2007	$167.5 - 163.7 = 3.8$

The largest increase was 2003 to 2004.

c. Answers will vary.
An estimated increase is 3.6 million, making the number of users in 2008: $167.5 + 3.6 = 171.1$ million.

Section 3.3

1.

0.7	*1 decimal place*
× 0.4	*1 decimal place*
0.28	*2 decimal places*

3.

0.07	*2 decimal places*
× 0.4	*1 decimal place*
0.028	*3 decimal places*

5.

0.03	*2 decimal places*
× 0.09	*2 decimal places*
0.0027	*4 decimal places*

7.

2.6	*1 decimal place*
× 0.3	*1 decimal place*
0.78	*2 decimal places*

9.

0.9	*1 decimal place*
× 0.88	*2 decimal places*
72	
720	
0.792	*3 decimal places*

11.

3.12	*2 decimal places*
× 0.005	*3 decimal places*
0.01560	*5 decimal places*

The trailing zero can be dropped: 0.0156

13.

4.003	*3 decimal places*
× 6.07	*2 decimal places*
28021	
2401800	
24.29821	*5 decimal places*

15.

0.006	*3 decimal places*
× 5	*0 decimal places*
0.030	*3 decimal places*

17.

75.14	*2 decimal places*
× 2.5	*1 decimal place*
37570	
150280	
187.850	*3 decimal places*

The trailing zero can be dropped: 187.85

19.

0.1	*1 decimal place*
× 0.02	*2 decimal places*
0.002	*3 decimal places*

21.

2.796	*3 decimal places*
× 10	*0 decimal places*
27.960	*3 decimal places*

The trailing zero can be dropped, and
$2.796(10) = 27.96$
(Shortcut for multiplying by 1<u>0</u> is to move decimal point 1 place to the right)

23.

0.0043	*4 decimal places*
× 100	*0 decimal places*
0.4300	*4 decimal places*

The trailing zero can be dropped: 0.43
(Shortcut for multiplying by 1<u>00</u> is to move decimal point 2 places to the right)

25.

49.94	*2 decimal places*
×1,000	*0 decimal places*
49,940.00	*2 decimal places*

The trailing zeros can be dropped: 49,940
(Shortcut for multiplying by 1<u>000</u> is to move decimal point 3 places to the right)

27.

987.654	*3 decimal places*
×10,000	*0 decimal places*
9,876,540.000	*3 decimal places*

The trailing zeros can be dropped:
9,876,540
(Shortcut for multiplying by 1<u>0,000</u> is to move decimal point 4 places to the right)

29. $2.1(3.5 - 2.6) = 2.1(0.9)$

$$= 1.89$$

31. $0.05(0.02 + 0.03) = 0.05(0.05)$

$$= 0.0025$$

33. $2.02(0.03 + 2.5) = 2.02(2.53)$

$$= 5.1106$$

35. $(2.1 + 0.03)(3.4 + 0.05) = (2.13)(3.45)$

$$= 7.3485$$

37. $(2.1 - 0.1)(2.1 + 0.1) = (2.0)(2.2)$

$$= 4.4$$

39. $3.08 - 0.2(5 + 0.03) = 3.08 - 0.2(5.03)$

$$= 3.08 - 1.006$$

$$= 2.074$$

41. $4.23 - 5(0.04 + 0.09) = 4.23 - 5(0.13)$

$$= 4.23 - 0.65$$

$$= 3.58$$

43. $2.5 + 10(4.3)^2 = 2.5 + 10(18.49)$

$$= 2.5 + 184.9$$

$$= 187.4$$

45. $100(1 + 0.08)^2 = 100(1.08)^2$

$$= 100(1.1664)$$

$$= 116.64$$

47. $(1.5)^2 + (2.5)^2 + (3.5)^2$

$$= (2.25) + (6.25) + (12.25)$$

$$= 20.75$$

Applying the Concepts

49. Product of 6 and the sum of 0.001 and 0.02

$$6 \times \qquad (0.001 + 0.02)$$

$$6(0.001 + 0.02) = 6(0.021)$$

$$= 0.126$$

51. For example:

1.2	0.12	0.012
× 100	× 100	× 100
$120.0 = 120$	$12.00 = 12$	$1.200 = 1.2$

When multiplying by 100, the decimal point moves 2 places to the right.

53.

Cost per $1,000:	$9.66
Number of $1,000's :	× 143
Total payment:	2898
	38640
	96600
	1,381.38

The monthly payment is $1,381.38

55.

$$\text{Cost of the first minute} + \text{Cost of the next 19 minutes} = \text{Total Cost}$$

$$\$0.45 \qquad + \qquad \$0.35(19)$$

$$0.45 + 0.35(19) = 0.45 + 6.65$$

$$= 7.10$$

The total cost is $7.10

57.

$$\text{Cost for 2 days} + \text{Cost for 120 miles}$$

$$\$15(2) \qquad + \qquad \$0.12(120)$$

$$15(2) + 0.12(120) = 30 + 14.4$$

$$= 44.0$$

The total cost is $44.40

59. Pay for 1st 36 hours + Pay for next 9 hours

$\quad\quad\$5.92(36)\quad\quad+\quad\quad\$8.88(9)$

$\quad 5.92(36)+8.88(9)=213.12+79.92$

$\quad\quad\quad\quad\quad\quad\quad\quad\quad = 293.04$

His total pay is \$293.04

61. $A = lw$

$\quad = (67 \text{ mm})(127 \text{ mm})$

$\quad = 8{,}509 \text{ mm}^2$

63. $A = lw$

$\quad = (1.41 \text{ in})(0.84 \text{ in})$

$\quad = 1.1844 \text{ in}^2$

$\quad \approx 1.18 \text{ in}^2$

65. $C = 2\pi r \quad\quad\quad A = \pi r^2$

$\quad = 2(3.14)(4 \text{ in}) \quad = (3.14)(4 \text{ in})^2$

$\quad = 25.12 \text{ in} \quad\quad = (3.14)(16 \text{ in}^2)$

$\quad\quad\quad\quad\quad\quad\quad = 50.24 \text{ in}^2$

67. $C = 2\pi r$

$\quad = 2(3.14)(3{,}900 \text{ mi})$

$\quad = 24{,}492 \text{ mi}$

69. Find the circumference:

$\quad C = 2\pi r$

$\quad = 2(3.14)(26.75 \text{ in})$

$\quad = 167.99 \text{ in}$

$\quad \approx 168 \text{ in}$

71. $V = \pi r^2 h$

$\quad = 3.14(2 \text{ ft})^2 (8 \text{ ft})$

$\quad = 3.14(4 \text{ ft}^2)(8 \text{ ft})$

$\quad = 100.48 \text{ ft}^3$

73. $V = \pi r^2 h$

$\quad = 3.14(2 \text{ ft})^2 (4 \text{ ft})$

$\quad = 3.14(4 \text{ ft}^2)(4 \text{ ft})$

$\quad = 50.24 \text{ ft}^3$

75. Note the radius is 1/2 of the diameter:

$\quad r = \dfrac{1}{2}d = \dfrac{1}{2}(26.5 \text{ mm}) = 13.25 \text{ mm}$

a. $C = 2\pi r$

$\quad = 2(3.14)(13.25 \text{ mm})$

$\quad = 83.21 \text{ mm}$

b. $A = \pi r^2$

$\quad = (3.14)(13.25 \text{ mm})^2$

$\quad = (3.14)(175.5625 \text{ mm}^2)$

$\quad = 551.26625 \text{ mm}^2$

$\quad \approx 551.27 \text{ mm}^2$

c. $V = \pi r^2 h$

$\quad = 3.14(13.25 \text{ mm})^2 (2.00 \text{ mm})$

$\quad = 3.14(175.5625 \text{ mm}^2)(2.00 \text{ mm})$

$\quad = 1{,}102.5325 \text{ mm}^3$

$\quad \approx 1{,}102.53 \text{ mm}^3$

Review Problems

77.

$$
\begin{array}{r}
1{,}879 \\
2\overline{)3{,}758} \\
\underline{2\downarrow} \\
17 \\
\underline{16\downarrow} \\
15 \\
\underline{14\downarrow} \\
18 \\
\underline{18} \\
0
\end{array}
$$

79.

$$\begin{array}{r} 1{,}516 \text{ R4} \\ 33\overline{)50{,}032} \end{array}$$

$$\begin{array}{r}
33\downarrow \\ \hline
170 \\
165\downarrow \\ \hline
53 \\
33\downarrow \\ \hline
202 \\
198 \\ \hline
4
\end{array}$$

Improving your Quantitative Literacy

81. Volume of the center tank:

$$V = \pi r^2 h$$
$$= 3.14(6 \text{ ft})^2 (16 \text{ ft})$$
$$= 3.14(36 \text{ ft}^2)(16 \text{ ft})$$
$$= 1{,}808.64 \text{ ft}^3$$

Volume of the containment tank:

$$V = \pi r^2 h$$
$$= 3.14(8 \text{ ft})^2 (4 \text{ ft})$$
$$= 3.14(64 \text{ ft}^2)(4 \text{ ft})$$
$$= 803.84 \text{ ft}^3$$

No. The containment tank holds less.

Section 3.4

1.
$$
\begin{array}{r}
19.7 \\
20\overline{)394.0} \\
\underline{20\downarrow} \\
194 \\
\underline{180\downarrow} \\
140 \\
\underline{140} \\
0
\end{array}
$$

3.
$$
\begin{array}{r}
6.2 \\
40\overline{)248.0} \\
\underline{240\downarrow} \\
80 \\
\underline{80} \\
0
\end{array}
$$

5.
$$
\begin{array}{r}
5.2 \\
5\overline{)26.0} \\
\underline{25\downarrow} \\
10 \\
\underline{10} \\
0
\end{array}
$$

7.
$$
\begin{array}{r}
11.04 \\
25\overline{)276.00} \\
\underline{25\downarrow} \\
26 \\
\underline{25\downarrow\downarrow} \\
100 \\
\underline{100} \\
0
\end{array}
$$

9.
$$
\begin{array}{r}
4.8 \\
6\overline{)28.8} \\
\underline{24\downarrow} \\
48 \\
\underline{48} \\
0
\end{array}
$$

11.
$$
\begin{array}{r}
9.7 \\
8\overline{)77.6} \\
\underline{72\downarrow} \\
56 \\
\underline{56} \\
0
\end{array}
$$

13.
$$
\begin{array}{r}
2.63 \\
35\overline{)92.05} \\
\underline{70\downarrow} \\
220 \\
\underline{210\downarrow} \\
105 \\
\underline{105} \\
0
\end{array}
$$

15.
$$
\begin{array}{r}
4.24 \\
45\overline{)190.80} \\
\underline{180\downarrow} \\
108 \\
\underline{90\downarrow} \\
180 \\
\underline{180} \\
0
\end{array}
$$

17.
$$34\overline{)86.70}$$ quotient 2.55

$$
\begin{array}{r}
2.55 \\
34\overline{)86.70} \\
\underline{68}\downarrow \\
187 \\
\underline{170}\downarrow \\
170 \\
\underline{170} \\
0
\end{array}
$$

19.
$$
\begin{array}{r}
1.35 \\
22\overline{)29.70} \\
\underline{22}\downarrow \\
77 \\
\underline{66}\downarrow \\
110 \\
\underline{110} \\
0
\end{array}
$$

21. $4.5\overline{)29.25} \rightarrow$
$$
\begin{array}{r}
6.5 \\
45\overline{)292.5} \\
\underline{270}\downarrow \\
225 \\
\underline{225} \\
0
\end{array}
$$

23. $0.11\overline{)1.089} \rightarrow$
$$
\begin{array}{r}
9.9 \\
11\overline{)108.9} \\
\underline{99}\downarrow \\
99 \\
\underline{99} \\
0
\end{array}
$$

25. $2.3\overline{)0.115} \rightarrow$
$$
\begin{array}{r}
0.05 \\
23\overline{)1.15} \\
\underline{115} \\
0
\end{array}
$$

27. $0.012\overline{)1.068} \rightarrow$
$$
\begin{array}{r}
89 \\
12\overline{)1,068} \\
\underline{96}\downarrow \\
108 \\
\underline{108} \\
0
\end{array}
$$

29. $1.1\overline{)2.42} \rightarrow$
$$
\begin{array}{r}
2.2 \\
11\overline{)24.2} \\
\underline{22}\downarrow \\
22 \\
\underline{22} \\
0
\end{array}
$$

31.
$$
\begin{array}{r}
1.346 \\
26\overline{)35.000} \\
\underline{26}\downarrow \\
90 \\
\underline{78}\downarrow \\
120 \\
\underline{104}\downarrow \\
160 \\
\underline{156} \\
4
\end{array}
$$

Rounding to the nearest hundredth: 1.35

33. $3.3\overline{)56} \rightarrow 33\overline{)560.000}$ quotient 16.969

```
        16.969
   33)560.000
       33↓
       230
       198↓
       320
       297↓
       230
       198
       320
       297
        23
```

Rounding to the nearest hundredth: 16.97

35. $0.5\overline{)0.1234} \rightarrow 5\overline{)1.234}$ quotient 0.246

```
       0.246
    5)1.234
      10↓
      23
      20↓
      34
      32
       2
```

Rounding to the nearest hundredth: 0.25

37. $7\overline{)19.000}$ quotient 2.714

```
      2.714
   7)19.000
     14↓
     50
     49↓
     10
      7↓
     30
     28
      2
```

Rounding the to nearest hundredth: 2.71

39. $0.059\overline{)0.69} \rightarrow 59\overline{)690.000}$ quotient 11.694

```
        11.694
   59)690.000
       59↓
       100
        59↓
       410
       354↓
       560
       531↓
       290
       236
        54
```

Rounding the to nearest hundredth: 11.69

41. $0.3\overline{)6.99} \rightarrow 3\overline{)69.9}$

False.

43. $0.06\overline{)42} \rightarrow 0.6\overline{)420} \rightarrow 6\overline{)4,200}$

True.

Applying the Concepts

45. $12 \ \cancel{\text{km}}\left(\dfrac{1 \text{ mi}}{1.61 \ \cancel{\text{km}}}\right) = \dfrac{12}{1.61} \text{ mi}$

$1.61\overline{)12.0} \rightarrow 161\overline{)1200.00}$ quotient 7.45

```
          7.45
  161)1200.00
      1127↓
       730
       644↓
       860
       805
        55
```

1.61 km is approximately 7.5 miles.

47. For all players:

$$\frac{\text{Total earnings}}{\text{Tournaments}} = \text{Average earnings}$$

Web: $\dfrac{\$1,591,959}{25} = \$63,678$

Inkster: $\dfrac{\$1,337,256}{24} = \$55,719$

Pak: $\dfrac{\$959,926}{27} = \$35,553$

Sorenstam: $\dfrac{\$863,816}{22} = \$39,264$

Kane: $\dfrac{\$757,844}{30} = \$25,261$

49. $\dfrac{\text{Total wages}}{\text{hours worked}} = \text{wage per hour}$

$\dfrac{\$33.90}{6} = \$5.65$

She earns $5.65 per hour.

51. $\dfrac{\text{Total miles}}{\text{gallons used}} = \text{miles per 1 gallon}$

$\dfrac{336 \text{ miles}}{15 \text{ gallons}} = 22.4 \text{ miles per gallon}$

53. First, find the amount of her total wages that came from overtime:

Total wage $-$ regular wage $=$ overtime wage

$\$294.93 \quad - \quad 36(6.78) \quad = 294.93 - 244.08$

$= 50.85$

$\dfrac{\text{Overtime wages}}{\text{Overtime wage per hour}} = \text{Overtime hours}$

$\dfrac{\$50.85}{\$10.17 \text{ per hour}} = 5 \text{ hours}$

She worked 5 hours overtime.

55. First find the amount of the call that comes from additional minutes:

Total cost of call: $\$2.33$

$-$ cost of 1st minute: $- \ 0.41$

Total cost of additional minutes: $\$1.92$

To find the number of additional minutes:

$\dfrac{\text{Total cost of additional minutes}}{\text{cost per minute}} = \dfrac{\$1.92}{\$0.32}$

$= 6$

The call was 7 minutes long (the 1st minute plus 6 additional minutes).

57. Complete the table as follows:

Class	Units		Value	Grade Points
Math	3	A	4	$3 \times 4 = 12$
Health	2	B	3	$2 \times 3 = 6$
History	3	B	3	$3 \times 3 = 9$
English	3	C	2	$3 \times 2 = 6$
Chemistry	4	C	2	$4 \times 2 = 8$
Total	15			41

$\dfrac{\text{Total grade points}}{\text{Total units}} = \text{GPA}$

$\dfrac{41}{15} = 2.73$

His GPA is 2.73

59. Complete the table as follows:

Class	Units		Value	Grade Points
Math	3	A	4	$3 \times 4 = 12$
Health	2	C	2	$2 \times 2 = 4$
History	3	B	3	$3 \times 3 = 9$
English	3	C	2	$3 \times 2 = 6$
Chemistry	4	C	2	$4 \times 2 = 8$
Total	15			39

$\dfrac{\text{Total grade points}}{\text{Total units}} = \text{GPA}$

$\dfrac{39}{15} = 2.6$

His GPA would have dropped from 2.73 to 2.6, a difference of 0.13

Calculator Exercises

61. $7 \div 9 = 0.77778$

63. $243 \div 0.791 = 307.20607$

65. $0.0503 \div 0.0709 = 0.70945$

Review Problems

67. $\dfrac{75}{100} = \dfrac{\cancel{25} \cdot 3}{\cancel{25} \cdot 4} = \dfrac{3}{4}$

69. $\dfrac{12}{18} = \dfrac{\cancel{6} \cdot 2}{\cancel{6} \cdot 3} = \dfrac{2}{3}$

71. $\dfrac{75}{100} = \dfrac{\cancel{25} \cdot 3}{\cancel{25} \cdot 4} = \dfrac{3}{4}$

73. $\dfrac{3}{5} = \dfrac{3 \cdot 2}{5 \cdot 2} = \dfrac{6}{10}$

75. $\dfrac{3}{5} = \dfrac{3 \cdot 20}{5 \cdot 20} = \dfrac{60}{100}$

77. $\dfrac{4}{5} = \dfrac{4 \cdot 3}{5 \cdot 3} = \dfrac{12}{15}$

79. $\dfrac{4}{1} = \dfrac{4 \cdot 15}{1 \cdot 15} = \dfrac{60}{15}$

81. $\dfrac{6}{5} = \dfrac{6 \cdot 3}{5 \cdot 3} = \dfrac{18}{15}$

Improving your Quantitative Literacy

83. a. No. Assume the original ages were given to the nearest year (a whole number). The median could either be a whole number, or possibly a decimal with decimal part equal to 0.5.

For example:
Ages: 18, 22, 27, 30, 33
Median: (middle value) 27

Ages: 18, 22, 27, 30, 33, 35
Median: (mean of the middle 2 values:)

$\text{Median} = \dfrac{27 + 30}{2} = \dfrac{57}{2} = 28.5$

b. The ages in the chart must be means.

Section 3.5

1. $\frac{1}{8}$ of the circle is shaded:

$$8\overline{)1.000} \qquad \frac{1}{8} = 0.125$$

0.125

$$\underline{8\downarrow}$$
$$20$$
$$\underline{16\downarrow}$$
$$40$$
$$\underline{40}$$
$$0$$

3. $\frac{5}{8}$ of the circle is shaded:

$$8\overline{)5.000} \qquad \frac{5}{8} = 0.625$$

0.625

$$\underline{48\downarrow}$$
$$20$$
$$\underline{16\downarrow}$$
$$40$$
$$\underline{40}$$
$$0$$

5.

0.25	0.5	0.75	1
$4\overline{)1.00}$	$4\overline{)2.0}$	$4\overline{)3.00}$	$4\overline{)4}$

0.25
$$4\overline{)1.00}$$
$$\underline{8\downarrow}$$
$$20$$
$$\underline{20}$$
$$0$$

0.5
$$4\overline{)2.0}$$
$$\underline{20}$$
$$0$$

0.75
$$4\overline{)3.00}$$
$$\underline{28\downarrow}$$
$$20$$
$$\underline{20}$$
$$0$$

1
$$4\overline{)4}$$
$$\underline{4}$$
$$0$$

Fraction	$\frac{1}{4}$	$\frac{2}{4}$	$\frac{3}{4}$	$\frac{4}{4}$
Decimal	0.25	0.5	0.75	1

7.

0.166
$$6\overline{)1.00}$$
$$\underline{6\downarrow}$$
$$40$$
$$\underline{36}$$
$$4$$

0.33
$$6\overline{)2.00}$$
$$\underline{18\downarrow}$$
$$20$$
$$\underline{18}$$
$$2$$

0.5
$$6\overline{)3.0}$$
$$\underline{30}$$
$$0$$

0.66
$$6\overline{)4.00}$$
$$\underline{36\downarrow}$$
$$40$$
$$\underline{36}$$
$$4$$

0.833
$$6\overline{)5.000}$$
$$\underline{48\downarrow}$$
$$20$$
$$\underline{18\downarrow}$$
$$20$$
$$\underline{18}$$
$$2$$

1
$$6\overline{)6}$$
$$\underline{6}$$
$$0$$

Fraction	$\frac{1}{6}$	$\frac{2}{6}$	$\frac{3}{6}$	$\frac{4}{6}$	$\frac{5}{6}$	$\frac{6}{6}$
Decimal	$0.1\overline{6}$	$0.\overline{3}$	0.5	$0.\overline{6}$	$0.8\overline{3}$	1

9.

0.48
$$25\overline{)12.00} \qquad \frac{12}{25} = 0.48$$
$$\underline{100\downarrow}$$
$$200$$
$$\underline{200}$$
$$0$$

11.

0.4375
$$32\overline{)14.0000} \qquad \frac{14}{32} = 0.4375$$
$$\underline{128\downarrow}$$
$$120$$
$$\underline{96\downarrow}$$
$$240$$
$$\underline{224\downarrow}$$
$$160$$
$$\underline{160}$$
$$0$$

13. $\begin{array}{r} 0.923 \\ 13\overline{)12.000} \\ \underline{117}\downarrow \\ 30 \\ \underline{26}\downarrow \\ 40 \\ \underline{39} \\ 1 \end{array}$ $\dfrac{12}{13} \approx 0.92$

15. $\begin{array}{r} 0.272 \\ 11\overline{)3.000} \\ \underline{22}\downarrow \\ 80 \\ \underline{77}\downarrow \\ 30 \\ \underline{22} \\ 8 \end{array}$ $\dfrac{3}{11} \approx 0.27$

17. $\begin{array}{r} 0.086 \\ 23\overline{)2.000} \\ \underline{184}\downarrow \\ 160 \\ \underline{138} \\ 22 \end{array}$ $\dfrac{3}{23} \approx 0.09$

19. $\begin{array}{r} 0.279 \\ 43\overline{)12.000} \\ \underline{86}\downarrow \\ 340 \\ \underline{301}\downarrow \\ 390 \\ \underline{387} \\ 3 \end{array}$ $\dfrac{12}{43} \approx 0.28$

21. The table is as follows:

Decimal	Fraction
0.125	$0.125 = \dfrac{125}{1,000} = \dfrac{1}{8}$
0.250	$0.250 = \dfrac{250}{1,000} = \dfrac{1}{4}$
0.375	$0.375 = \dfrac{375}{1,000} = \dfrac{3}{8}$
0.500	$0.500 = \dfrac{500}{1,000} = \dfrac{1}{2}$
0.625	$0.625 = \dfrac{625}{1,000} = \dfrac{5}{8}$
0.750	$0.750 = \dfrac{750}{1,000} = \dfrac{3}{4}$
0.875	$0.875 = \dfrac{875}{1,000} = \dfrac{7}{8}$

23. $0.15 = \dfrac{15}{100} = \dfrac{3}{20}$

25. $0.08 = \dfrac{8}{100} = \dfrac{2}{25}$

27. $0.375 = \dfrac{375}{1,000} = \dfrac{3}{8}$

29. $5.6 = 5\dfrac{6}{10} = 5\dfrac{3}{5}$

31. $5.06 = 5\dfrac{6}{100} = 5\dfrac{3}{50}$

33. $1.22 = 1\dfrac{22}{100} = 1\dfrac{11}{50}$

35. $\dfrac{1}{2}(2.3 + 2.5) = \dfrac{1}{2}(4.8)$
$= 2.4$

37. $\frac{3}{8}(4.7) = \frac{3(4.7)}{8}$

$\qquad = \frac{14.1}{8}$

$\qquad = 1.7625$

39. $3.4 - \frac{1}{2}(0.76) = 3.4 - 0.38$

$\qquad = 3.02$

41. $\frac{2}{5}(0.3) + \frac{3}{5}(0.3) = 0.4(0.3) + 0.6(0.3)$

$\qquad = 0.12 + 0.18$

$\qquad = 0.3$

43. $6\left(\frac{3}{5}\right)(0.02) = 6(0.6)(0.02)$

$\qquad = 0.072$

45. $\frac{5}{8} + 0.35\left(\frac{1}{2}\right) = \frac{5}{8} + 0.175$

$\qquad = 0.625 + 0.175$

$\qquad = 0.8$

47. $\left(\frac{1}{3}\right)^2 (5.4) + \left(\frac{1}{2}\right)^3 (3.2)$

$\quad = \left(\frac{1}{9}\right)(5.4) + \left(\frac{1}{8}\right)(3.2)$

$\quad = 0.6 + 0.4$

$\quad = 1$

49. $(0.25)^2 + \left(\frac{1}{4}\right)^2 (3) = 0.0625 + \left(\frac{1}{16}\right)(3)$

$\qquad = 0.0625 + 0.1875$

$\qquad = 0.25$

Applying the Concepts

51.

Cost per pound	×	Number of pounds	=	Total Cost
$2.59	×	$3\frac{1}{4}$	=	$2.59(3.25)$

$\qquad\qquad\qquad\qquad = 8.4175$

Rounding to the nearest cent, the cost is $8.42.

53. Markdown $= \frac{1}{3}(\$57.99) = \$19.33$

Sale price = Original price − Markdown

$\qquad = \$57.99 - \$19.33$

$\qquad = \$38.66$

55. Find the perimeter of the shaded region by summing the perimeters of the 3 small shaded triangles:

$3P = 3(1 \text{ in} + 1 \text{ in} + 1 \text{ in}) = 3(3 \text{ in}) = 9 \text{ in}$

57.

$$\begin{array}{r} 0.75 \\ 4\overline{)3.00} \\ \underline{28}\downarrow \\ 20 \\ \underline{20} \\ 0 \end{array}$$

$$\begin{array}{r} 0.562 \\ 16\overline{)9.000} \\ \underline{80}\downarrow \\ 100 \\ \underline{96}\downarrow \\ 40 \\ \underline{32} \\ 8 \end{array}$$

$$\begin{array}{r} 0.093 \\ 32\overline{)3.000} \\ \underline{288}\downarrow \\ 120 \\ \underline{96} \\ 24 \end{array}$$

$$\begin{array}{r} 0.218 \\ 32\overline{)7.000} \\ \underline{64}\downarrow \\ 60 \\ \underline{32}\downarrow \\ 280 \\ \underline{256} \\ 24 \end{array}$$

$$\begin{array}{r} 0.062 \\ 16\overline{)1.000} \\ \underline{96}\downarrow \\ 40 \\ \underline{32} \\ 8 \end{array}$$

Date	Gain/Loss($)	Decimal
March 6	$\dfrac{3}{4}$	0.75
March 7	$\dfrac{9}{16}$	0.56
March 8	$\dfrac{3}{32}$	0.09
March 9	$\dfrac{7}{32}$	0.22
March 10	$\dfrac{1}{16}$	0.06

59. Calories in ground beef $= 50.75\left(4\dfrac{1}{2}\right)$

$$= 50.75(4.5)$$
$$= 228.375$$

Calories in halibut $= 27.5\left(4\dfrac{1}{2}\right)$

$$= 27.5(4.5)$$
$$= 123.75$$

Difference in calories $= 228.375 - 123.75$
$$= 104.625 \text{ cal}$$

61. $V = \dfrac{4}{3}\pi r^3$

$$= \dfrac{4}{3}(3.14)(2 \text{ mi})^3$$
$$= \dfrac{4}{3}(3.14)(8 \text{ mi}^3)$$
$$= \dfrac{4}{3}(25.12)\text{mi}^3$$
$$= 33.49\overline{3} \text{ mi}^3$$

63. $V = \dfrac{4}{3}\pi r^3$

$$= \dfrac{4}{3}(3.14)(3.9 \text{ in})^3$$
$$= \dfrac{4}{3}(3.14)(59.319\,\text{in}^3)$$
$$= \dfrac{4}{3}(186.26166 \text{ in}^3)$$
$$= 248.34888 \text{ in}^3$$

The volume is 248.35 in^3.

65.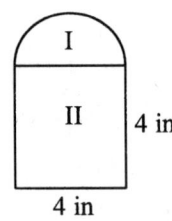

Separate into 2 regions. Note the radius of

the circle is $\frac{1}{2}(\text{diameter}) = \frac{1}{2}(4 \text{ in}) = 2 \text{ in}$.

Also, the area of *half* a circle is $\frac{1}{2}(\pi r^2)$.

Area = Area I + Area II

$= \frac{1}{2}\pi r^2 + lw$

$= \frac{1}{2}(3.14)(2)^2 + 4(4)$

$= \frac{1}{2}(3.14)(4) + 16$

$= 6.28 + 16$

$= 22.28$

The total area is 22.28 in^2

67. Separate into 2 solids (a cylinder and hemisphere). Note that the volume of *half* a sphere is $\frac{1}{2}\left(\frac{4}{3}\pi r^3\right)$

$V = \text{Cylinder volume} + \text{hemisphere volume}$

$= \pi r^2 h + \frac{1}{2}\left(\frac{4}{3}\pi r^3\right)$

$= 3.14(3)^2(6) + \frac{1}{2}\left(\frac{4}{3}\right)(3.14)(3)^3$

$= 3.14(9)(6) + \frac{2}{3}(3.14)(27)$

$= 169.56 + 56.52$

$= 226.08$

The volume is 226.1 ft^3

Calculator Problems

69. $\frac{7}{9} = 0.\overline{7}$

71. $\frac{123}{999} = 0.\overline{123}$

73. Convert $\frac{1}{5}$ mi $= 0.2$ mi.

To calculate the costs after the first 0.2 miles, note that \$0.25 for $\frac{1}{5}$ of a mile is equivalent to \$1.25 for 1 mile (multiply both quantities by 5).

$\underbrace{\$12.5 \text{ for}}_{\text{1st 0.2 miles}} + \underbrace{\$1.25 \text{ per mile for}}_{\text{additional miles}} = \text{Total Cost}$

$\underbrace{\$1.25}_{} + \underbrace{\$1.25(7.5 - 0.2)}_{}$

$1.25 + 1.25(7.5 - 0.2) = 1.25 + 1.25(7.3)$

$= 1.25 + 9.125$

$= 10.375$

The total cost is \$10.38

75. See discussion above in problem #73.

$12.5 for + $1.25 per mile for = Total

$\underbrace{\text{1st 0.2 miles}}$ $\underbrace{\text{additional miles}}$ Cost

$$\$1.25 \quad + \$1.25(12.4-0.2)$$
$$1.25 + 1.25(12.4-0.2) = 1.25 + 1.25(12.2)$$
$$= 1.25 + 15.25$$
$$= 16.50$$

The total cost is $16.50. The meter is correct.

Review Problems

77. $5^3 = 5 \cdot 5 \cdot 5 = 125$

79. $3^2 = 3 \cdot 3 = 9$

81. $\left(\dfrac{1}{3}\right)^4 = \left(\dfrac{1}{3}\right)\left(\dfrac{1}{3}\right)\left(\dfrac{1}{3}\right)\left(\dfrac{1}{3}\right) = \dfrac{1}{81}$

83. $\left(\dfrac{5}{6}\right)^2 = \left(\dfrac{5}{6}\right)\left(\dfrac{5}{6}\right) = \dfrac{25}{36}$

85. $(0.5)^2 = (0.5)(0.5) = 0.25$

87. $1.2^2 = (1.2)(1.2) = 1.44$

Improving your Quantitative Literacy

89. a. iTunes $< \dfrac{1}{3}$(Walmart)

$$\$9.99 < \dfrac{1}{3}(\$27.28)$$
$$\$9.99 < \$9.09$$
false

b. Walmart > 3(iTunes)
$$\$27.28 > 3(\$9.99)$$
$$\$27.28 > \$29.97$$
false

c. Walmart is < 3(iTunes)
$$\$27.28 < 3(\$9.99)$$
$$\$27.28 < \$29.97$$
true

Section 3.6

1. $\sqrt{64} = 8$, because $8^2 = 64$

3. $\sqrt{81} = 9$, because $9^2 = 81$

5. $\sqrt{36} = 6$, because $6^2 = 36$

7. $\sqrt{25} = 5$, because $5^2 = 25$

9. $3\sqrt{25} = 3 \cdot 5 = 15$

10. $9\sqrt{49} = 9 \cdot 7 = 63$

11. $6\sqrt{64} = 6 \cdot 8 = 48$

13. $15\sqrt{9} = 15 \cdot 3 = 45$

15. $16\sqrt{9} = 16 \cdot 3 = 48$

17. $\sqrt{49} + \sqrt{64} = 7 + 8$
$= 15$

19. $\sqrt{16} - \sqrt{9} = 4 - 3$
$= 1$

21. $3\sqrt{25} + 9\sqrt{49} = 3 \cdot 5 + 9 \cdot 7$
$= 15 + 63$
$= 78$

23. $15\sqrt{9} - 9\sqrt{16} = 15 \cdot 3 - 9 \cdot 4$
$= 45 - 36$
$= 9$

25. $\sqrt{\dfrac{16}{49}} = \dfrac{4}{7}$, because $\left(\dfrac{4}{7}\right)^2 = \dfrac{16}{49}$

27. $\sqrt{\dfrac{36}{64}} = \dfrac{6}{8} = \dfrac{3}{4}$, because $\left(\dfrac{6}{8}\right)^2 = \dfrac{36}{64}$

29. Work each side of the equation separately:
$$\sqrt{4} + \sqrt{9} = \sqrt{4+9}$$
$$2 + 3 = \sqrt{13}$$
$$5 = \sqrt{13} \quad \text{false}$$

31. Work each side of the equation separately:
$$\sqrt{25 \cdot 9} = \sqrt{25} \cdot \sqrt{9}$$
$$\sqrt{225} = 5 \cdot 3$$
$$15 = 15 \quad \text{true}$$

33. $c = \sqrt{a^2 + b^2}$
$= \sqrt{6^2 + 8^2}$
$= \sqrt{36 + 64}$
$= \sqrt{100}$
$= 10$
$c = 10$ in.

35. $c = \sqrt{a^2 + b^2}$
$= \sqrt{5^2 + 12^2}$
$= \sqrt{25 + 144}$
$= \sqrt{169}$
$= 13$
$c = 13$ ft.

37. $c = \sqrt{a^2 + b^2}$
$= \sqrt{4^2 + 5^2}$
$= \sqrt{16 + 25}$
$= \sqrt{41}$
≈ 6.403124
$c = 6.40$ in.

39. $c = \sqrt{a^2 + b^2}$
$= \sqrt{9^2 + 15^2}$
$= \sqrt{81 + 225}$
$= \sqrt{306}$
≈ 17.492856
$c = 17.49$ m.

Applying the Concepts

41. Let c = length of wire (ft)

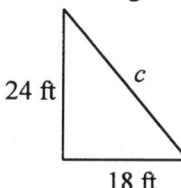

24 ft c

18 ft

$c = \sqrt{a^2 + b^2}$
$= \sqrt{24^2 + 18^2}$
$= \sqrt{576 + 324}$
$= \sqrt{900}$
$= 30$
The wire is 30 feet long.

43. Make a sketch. Label all known and unknown quantities.
Let c = length of ladder (ft)

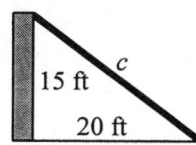

15 ft c

20 ft

$c = \sqrt{a^2 + b^2}$
$= \sqrt{15^2 + 20^2}$
$= \sqrt{225 + 400}$
$= \sqrt{625}$
$= 25$
The ladder is 25 feet long.

45. See "The Spiral of Roots" in chapter introduction.

Calculator Problems

47. $\sqrt{1.25} \approx 1.118034$
$= 1.1180$ (to nearest ten thousandth)

49. $\sqrt{125} \approx 11.180340$
$= 11.1803$ (to nearest ten thousandth)

51. $2\sqrt{3} \approx 2(1.732051)$
$= 3.464102$
$= 3.46$ (to nearest hundredth)

53. $5\sqrt{5} \approx 5(2.236068)$
$= 11.18034$
$= 11.18$ (to nearest hundredth)

55. $\dfrac{\sqrt{3}}{3} \approx \dfrac{1.732051}{3}$
≈ 0.577350
$= 0.58$ (to nearest hundredth)

57. $\sqrt{\dfrac{1}{3}} \approx \sqrt{0.333333}$
≈ 0.577340
$= 0.58$ (to nearest hundredth)

59. $\sqrt{12} + \sqrt{75} \approx 3.464102 + 8.660254$
$= 12.124356$
$= 12.124$ (to nearest thousandth)

61. $\sqrt{87} \approx 9.327379$
$= 9.327$ (to nearest thousandth)

63. $2\sqrt{3}+5\sqrt{3}\approx 2(1.732051)+5(1.732051)$

$\qquad = 3.464102+8.660255$

$\qquad = 12.124357$

$\qquad = 12.124$ (to nearest thousandth)

65. $7\sqrt{3}\approx 7(1.732051)$

$\qquad = 12.124357$

$\qquad = 12.124$ (to nearest thousandth)

67. The table is as follows:

Height, h (ft)	Distance, d (miles)
10	$d=\sqrt{\dfrac{3\cdot 10}{2}}=\sqrt{15}\approx 4$
50	$d=\sqrt{\dfrac{3\cdot 50}{2}}=\sqrt{75}\approx 9$
90	$d=\sqrt{\dfrac{3\cdot 90}{2}}=\sqrt{135}\approx 12$
130	$d=\sqrt{\dfrac{3\cdot 130}{2}}=\sqrt{195}\approx 14$
170	$d=\sqrt{\dfrac{3\cdot 170}{2}}=\sqrt{255}\approx 16$
190	$d=\sqrt{\dfrac{3\cdot 190}{2}}=\sqrt{285}\approx 17$

Review Problems

69. $\dfrac{5}{7}\cdot\dfrac{14}{25}=\dfrac{\cancel{5}\cdot 2\cdot\cancel{7}}{\cancel{7}\cdot\cancel{5}\cdot 5}=\dfrac{2}{5}$

71. $4\dfrac{3}{10}=4\dfrac{3\cdot 10}{10\cdot 10}=4\dfrac{30}{100}$

$+\;5\dfrac{2}{100}=+\;5\dfrac{2}{100}=+\;5\dfrac{2}{100}$

$\qquad\qquad\qquad\qquad\qquad 9\dfrac{32}{100}=9\dfrac{8}{25}$

73. $3\dfrac{2}{10}\cdot 2\dfrac{5}{10}=\dfrac{32}{10}\cdot\dfrac{25}{10}=\dfrac{\cancel{2}\cdot\cancel{2}\cdot 8\cdot\cancel{5}\cdot\cancel{5}}{\cancel{2}\cdot\cancel{5}\cdot\cancel{2}\cdot\cancel{5}}=8$

75. $7\dfrac{1}{10}=\;6\dfrac{11}{10}$

$-\,4\dfrac{3}{10}=-\,4\dfrac{3}{10}$

$\qquad\qquad 2\dfrac{8}{10}=2\dfrac{4}{5}$

Extending the Concepts

77.

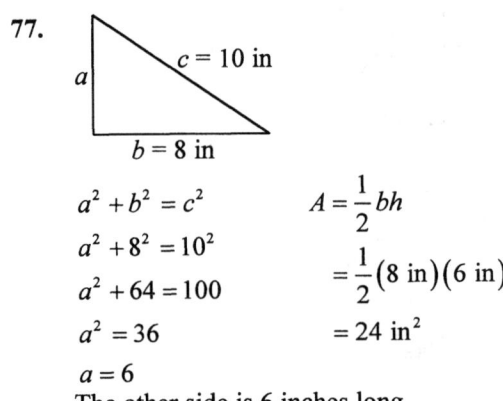

$a^2+b^2=c^2$ $\qquad\qquad A=\dfrac{1}{2}bh$

$a^2+8^2=10^2$

$a^2+64=100$ $\qquad\qquad =\dfrac{1}{2}(8\text{ in})(6\text{ in})$

$a^2=36$ $\qquad\qquad\qquad =24\text{ in}^2$

$a=6$

The other side is 6 inches long.

79.

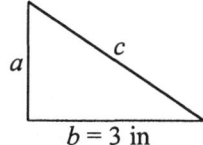

First find the height using the area formula:

$$A = \frac{1}{2}bh$$

$$6 = \frac{1}{2}(3)h$$

$$6 = \frac{3}{2}h$$

$$\frac{2}{3}(6) = \frac{2}{3}\left(\frac{3}{2}h\right)$$

$$h = 4 \text{ ft}$$

$$c = \sqrt{a^2 + b^2}$$
$$ = \sqrt{4^2 + 3^2}$$
$$ = \sqrt{16 + 9}$$
$$ = \sqrt{25}$$
$$ = 5$$

The hypotenuse is 5 feet long.

Improving Your Quantitative Literacy

81. a. Answers will vary. Look at the trend:
From 1971 to 1981: 3.7 student decrease
From 1981 to 1991: 1.7 student decrease
From 1991 to 2001: 1.2 student decrease
If the trend continues, there may be a 1.0
student decrease in from 2001 to 2011:
$15.8 - 1.0 = 14.8$

b. Answers will vary. Since 1996 is half
way between 1991 and 2001, you could
use the mean as an estimate:

$$\text{Mean} = \frac{17 + 15.8}{2} = 16.4 \text{ students}$$

Chapter 3 Review

1. 36.007

    ↑

    thousandths

3. Thirty-seven and forty-two ten thousandths is 37.0042

5. Look at the digit to the right of the hundredths place: 98.7654
$5 \geq 5$ so add 1 to the hundredths place and delete all digits to the right   98.77

7.
$$\begin{array}{r} 11.076 \\ - \ 3.297 \\ \hline 7.779 \end{array}$$

9.
$$\begin{array}{r} 24.24 \ \textit{2 decimal places} \\ \times \ 0.89 \ \textit{2 decimal places} \\ \hline 21816 \\ 193920 \\ \hline 21.5736 \ \textit{4 decimal places} \end{array}$$

11. $0.58\overline{)0.7134} \rightarrow 58\overline{)71.34}$

$$\begin{array}{r} 1.23 \\ 58\overline{)71.34} \\ \underline{58\downarrow} \\ 133 \\ \underline{116\downarrow} \\ 174 \\ \underline{174} \\ 0 \end{array}$$

13. $0.705 = \dfrac{705}{1,000} = \dfrac{141}{200}$

15. $3.3 - 4(0.22) = 3.3 - 0.88$
$$= 2.42$$

17. $125\left(\dfrac{3}{5}\right) + 4 = 75 + 4$
$$= 79$$

19. $3\sqrt{25} = 3 \cdot 5 = 15$

21. $4\sqrt{25} + 3\sqrt{81} = 4 \cdot 5 + 3 \cdot 9$
$$= 20 + 27$$
$$= 47$$

23. $P = 2l + 2w$
$$= 2(6.14 \text{ in}) + 2(2.56 \text{ in})$$
$$= 12.28 \text{ in} + 5.12 \text{ in}$$
$$= 17.4 \text{ in}$$

25. $r = \dfrac{d}{t} = \dfrac{3,106.3 \text{ mi}}{137.05 \text{ hr}} = 22.7\dfrac{\text{mi}}{\text{hr}}$

27a. Complete the table as follows:

Class	Units	Value		Grade Points
Algebra	4	C	2	$4 \times 2 = 8$
Speech	3	A	4	$3 \times 4 = 12$
Accting	3	B	3	$3 \times 3 = 9$
Mrktng	3	B	3	$3 \times 3 = 9$
Real Estate	2	C	2	$2 \times 2 = 4$
Total	15			42

$$\frac{\text{Total grade points}}{\text{Total units}} = \text{GPA}$$

$$\frac{42}{15} = 2.80$$

Her GPA is 2.80.

27b. The table is as follows:

Class	Units	Value		Grade Points
Algebra	4	B	3	$4 \times 3 = 12$
Speech	3	A	4	$3 \times 4 = 12$
Accting	3	B	3	$3 \times 3 = 9$
Markting	3	B	3	$3 \times 3 = 9$
Real Estate	2	C	2	$2 \times 2 = 4$
Total	15			46

$$\frac{\text{Total grade points}}{\text{Total units}} = \text{GPA}$$

$$\frac{46}{15} = 3.0\overline{6}$$

The difference between $3.0\overline{6}$ and 2.8
≈ 0.27

29.
$$c = \sqrt{a^2 + b^2}$$
$$= \sqrt{5^2 + 2^2}$$
$$= \sqrt{25 + 4}$$
$$= \sqrt{29}$$
$$\approx 5.4$$

The hypotenuse is 5.4 ft long.

31. $P = $ sum of all sides $\qquad A = \frac{1}{2}bh$

≈ 5 ft $+ 2$ ft $+ 5.4$ ft $\qquad = \frac{1}{2}(5 \text{ ft})(2 \text{ ft})$

$= 12.4$ ft $\qquad\qquad\qquad = \frac{1}{2}(10 \text{ ft}^2)$

$$= 5 \text{ ft}^2$$

33. Note: The radius is half the diameter:

$$r = \frac{1}{2}d = \frac{1}{2}(32 \text{ in}) = 16 \text{ in}.$$
$$C = 2\pi r$$
$$= 2(3.14)(16 \text{ in})$$
$$= 100.48 \text{ in}$$

He will need 100.48 inches of pipe.

Cumulative Review Chapters 1-3

1.
$$\begin{array}{r} 3,781 \\ +\ 298 \\ \hline 4,079 \end{array}$$

3.
$$\begin{array}{r} 287 \\ \times\ 56 \\ \hline 1,722 \\ 14,350 \\ \hline 16,072 \end{array}$$

5.
$$\begin{array}{r} 6.22 \\ 24\overline{)149.28} \\ 144\downarrow \\ \hline 52 \\ 48\downarrow \\ \hline 48 \\ 48 \\ \hline 0 \end{array}$$

7.
$$\begin{array}{r} 12.96 \\ \times\ 4.3 \\ \hline 3888 \\ 51840 \\ \hline 55.728 \end{array}$$

9. $\dfrac{5}{14} \div \dfrac{15}{21} = \dfrac{5}{14} \cdot \dfrac{21}{15} = \dfrac{\cancel{5} \cdot \cancel{3} \cdot \cancel{7}}{\cancel{7} \cdot 2 \cdot \cancel{3} \cdot \cancel{5}} = \dfrac{1}{2}$

11.
$$\begin{array}{r} 15 \\ 4\overline{)63} \\ 4\downarrow \\ \hline 23 \\ 20 \\ \hline 3 \end{array}$$
$\dfrac{63}{4} = 15 + \dfrac{3}{4} = 15\dfrac{3}{4}$

13. $2\dfrac{1}{2} \cdot 8 = \dfrac{5}{2} \cdot 8 = \dfrac{5 \cdot \cancel{2} \cdot 4}{\cancel{2}} = 20$

15. $\dfrac{72}{8} = 9$

17. $\underbrace{\text{Three times}}\ \underbrace{\text{the sum of 13 and 4}}\ \text{is } 51$

$\qquad 3\times \qquad\qquad (13+4) \qquad\qquad = 51$

$\qquad 3(13+4) = 51$

$\qquad\quad 3(17) = 51$

$\qquad\qquad 51 = 51$

19. False. For example,
$$\dfrac{1}{2} \neq \dfrac{1+1}{2+1} = \dfrac{1}{3}$$

21. $\dfrac{6+2(4)}{8+10} = \dfrac{6+8}{18} = \dfrac{14}{18} = \dfrac{7}{9}$

23. $\dfrac{2}{3}(0.45) + \dfrac{4}{5}(0.8) = \dfrac{2(0.45)}{3} + 0.8(0.8)$

$\qquad\qquad\qquad\qquad \dfrac{-0.90}{3} + 0.64$

$\qquad\qquad\qquad\quad = 0.30 + 0.64$

$\qquad\qquad\qquad\quad = 0.94$

25. $\left(3\dfrac{1}{3} - \dfrac{1}{2}\right)\left(4\dfrac{1}{2} + \dfrac{3}{4}\right) = \left(\dfrac{10}{3} - \dfrac{1}{2}\right)\left(\dfrac{9}{2} + \dfrac{3}{4}\right)$

$\qquad\qquad\qquad\qquad\qquad = \left(\dfrac{20}{6} - \dfrac{3}{6}\right)\left(\dfrac{18}{4} + \dfrac{3}{4}\right)$

$\qquad\qquad\qquad\qquad\qquad = \left(\dfrac{17}{6}\right)\left(\dfrac{21}{4}\right)$

$\qquad\qquad\qquad\qquad\qquad = \dfrac{17 \cdot \cancel{3} \cdot 7}{\cancel{3} \cdot 2 \cdot 4}$

$\qquad\qquad\qquad\qquad\qquad = \dfrac{119}{8}$

$\qquad\qquad\qquad\qquad\qquad = 14\dfrac{7}{8}$

27.

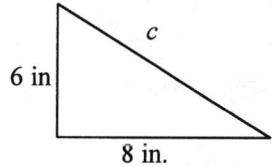

$$c = \sqrt{a^2 + b^2}$$
$$= \sqrt{6^2 + 8^2}$$
$$= \sqrt{36 + 64}$$
$$= \sqrt{100}$$
$$= 10 \text{ in.}$$

29. $V = lwh$
$$= (5 \text{ in})(3 \text{ in})(7 \text{ in})$$
$$= 105 \text{ in}^3$$

Chapter 3 Test

1. 5.053 in words:
Five and fifty-three thousandths

3. Seventeen and four hundred six ten
thousandths is 17.0406

5. Note: $7 = 7.00$

$$0.6 = 0.60$$

$$
\begin{array}{r}
7.00 \\
0.60 \\
+\ 0.58 \\
\hline
8.18
\end{array}
$$

7.

$$
\begin{array}{r}
6.24 \\
\times\ 5.7 \\
\hline
4368 \\
31200 \\
\hline
35.568
\end{array}
$$

9.

$$
\begin{array}{r}
0.92 \\
25\overline{)23.00} \\
225\downarrow \\
\hline
50 \\
50 \\
\hline
0
\end{array}
$$

11. $5.2(2.8 + 0.2) = 5.2(2.82)$

$$= 14.664$$

13. $23.852 - 3(2.01 + 0.231)$

$$= 23.852 - 3(2.241)$$

$$= 23.852 - 6.723$$

$$= 17.129$$

15. $2\sqrt{36} + 3\sqrt{64} = 2(6) + 3(8) = 12 + 24 = 36$

17.

$$
\begin{array}{lr}
\text{Money tendered:} & \$20.00 \\
-\text{ total purchases:} & -\ 8.47 \\
\hline
\text{Change:} & \$11.53
\end{array}
$$

19. $\dfrac{\$262}{40 \text{ hrs}} \rightarrow$

$$
\begin{array}{r}
6.55 \\
40\overline{)262.00} \\
240\downarrow \\
\hline
220 \\
200\downarrow \\
\hline
200 \\
200 \\
\hline
0
\end{array}
$$

$$\frac{\$262}{40 \text{ hrs}} = \frac{\$6.55}{\text{hr}}$$

21. a. The closest ages to 25 years is 24.9
years, which occurred in 2000.

b.

$$
\begin{array}{lr}
\text{Age in 1995:} & 24.5 \text{ years} \\
-\text{ Age in 1975:} & -\ 21.8 \text{ years} \\
\hline
\text{Increase:} & 2.7 \text{ years}
\end{array}
$$

c. Answers will vary. It appears the
average age is currently increasing at a
rate of 0.6 years of age for every 5 years.
If the trend continues:

Year	Age
2000	24.9
2005	$24.9 + 0.6 = 25.5$
2010	$25.5 + 0.6 = 26.1$

The age in 2010 may be about 26 years
old.

Chapter 3: A Glimpse of Algebra

1. $4x^2 + 2x + 3$
   $+\ \underline{2x^2 + 7x + 5}$
   $6x^2 + 9x + 8$

3. $2a + 3$
   $+\ \underline{3a + 5}$
   $5a + 8$

5. $3x + 4$
   $2x + 1$
   $+\ \underline{4x + 1}$
   $9x + 6$

7. $2y^3 + 3y^2 + 4y + 5$
   $+\ \underline{3y^3 + 2y^2 + 5y + 2}$
   $5y^3 + 5y^2 + 9y + 7$

9. $3x^2 + 4x + 3$
   $+\ \underline{3x^2 + 0x + 2}$
   $6x^2 + 4x + 5$

11. $3a^2 + 0a + 4$
    $+\ \underline{7a + 2}$
    $3a^2 + 7a + 6$

13. $5x^3 + 4x^2 + 7x + 3$
    $+\ \underline{3x^2 + 9x + 10}$
    $5x^3 + 7x^2 + 16x + 13$

Chapter 4 Preview

1. $\dfrac{16}{48} = \dfrac{\cancel{16}}{\cancel{16} \cdot 3} = \dfrac{1}{3}$

3. $\begin{array}{r} 0.25 \\ 4\overline{)1.00} \\ \underline{8\downarrow} \\ 20 \\ \underline{20} \\ 0 \end{array}$ $\dfrac{1}{4} = 0.25$

5. $3(0.4) = 1.2$

7. $\dfrac{2}{3} \cdot 6 = \dfrac{2 \cdot 2 \cdot \cancel{3}}{\cancel{3}} = 4$

9. $\begin{array}{r} 0.695 \\ 2\overline{)1.390} \\ \underline{12\downarrow} \\ 19 \\ \underline{18\downarrow} \\ 10 \\ \underline{10} \\ 0 \end{array}$

11. $\dfrac{\frac{2}{3}}{\frac{4}{9}} = \dfrac{2}{3} \cdot \dfrac{9}{4} = \dfrac{\cancel{2} \cdot \cancel{3} \cdot 3}{\cancel{3} \cdot \cancel{2} \cdot 2} = \dfrac{3}{2}$ or $1\dfrac{1}{2}$

13. $4 \cdot 6 = 3 \cdot 8$
    $24 = 24$
    true

15. $65 = y \cdot 10$
    $\dfrac{65}{10} = \dfrac{y \cdot 10}{10}$
    $6.5 = y$

Chapter 4 Pretest

1. $\dfrac{15}{25} = \dfrac{3}{5}$

3. $\dfrac{\frac{5}{4}}{\frac{7}{4}} = \dfrac{5}{4} \cdot \dfrac{4}{7} = \dfrac{5}{7}$

5. a. $\dfrac{\text{fees}}{\text{personal expenses}} = \dfrac{\$6,000}{\$1,600} = \dfrac{15}{4}$

 b. $\dfrac{\text{books/supplies}}{\text{fees}} = \dfrac{\$1,450}{\$6,000} = \dfrac{29}{120}$

7. $\dfrac{5 \text{ gallons}}{40 \text{ seconds}} = \dfrac{1}{8} \dfrac{\text{gal}}{\text{sec}} = 0.125 \text{ gal per sec}$

9. $\dfrac{399\cancel{c}}{12 \text{ cans}} = 33.25\cancel{c} \text{ per can}$

11. $\dfrac{2}{y} = \dfrac{4}{10}$

$2 \cdot 10 = y \cdot 4$

$20 = 4y$

$5 = y$

13. $\dfrac{n}{8} = \dfrac{\frac{1}{6}}{\frac{1}{9}}$

$n \cdot \dfrac{1}{9} = 8 \cdot \dfrac{1}{6}$

$\dfrac{n}{9} = \dfrac{8}{6}$

$9 \cdot \dfrac{n}{9} = 9 \cdot \dfrac{8}{6}$

$n = \dfrac{72}{6} = 12$

15. Let x = ml of essential oil

$\dfrac{7 \text{ ml oil}}{40 \text{ ml alcohol}} = \dfrac{x \text{ ml oil}}{20 \text{ ml alcohol}}$

$7(20) = 40x$

$140 = 40x$

$3.5 = x$

3.5 ml of essential oil.

17. Let x = amount reimbursed for 150 miles

$\dfrac{36.5\cancel{c}}{1 \text{ mi}} = \dfrac{x \ \cancel{c}}{150 \text{ mi}}$

$36.5(150) = 1(x)$

$5,475 = x$

They will be reimbursed 5,475¢, or \$54.75

19. $\dfrac{6}{15} = \dfrac{20}{BC}$

$6 \cdot BC = 300$

$BC = 50$

Section 4.1

1. $\dfrac{8}{6} = \dfrac{4}{3}$

3. $\dfrac{64}{12} = \dfrac{16}{3}$

5. $\dfrac{100}{250} = \dfrac{2}{5}$

7. $\dfrac{13}{26} = \dfrac{1}{2}$

9. $\dfrac{\frac{3}{4}}{\frac{1}{4}} = \dfrac{3}{4} \cdot \dfrac{4}{1} = \dfrac{3}{1}$

11. $\dfrac{\frac{7}{3}}{\frac{6}{3}} = \dfrac{7}{3} \cdot \dfrac{3}{6} = \dfrac{7}{6}$

13. $\dfrac{\frac{6}{5}}{\frac{6}{7}} = \dfrac{6}{5} \cdot \dfrac{7}{6} = \dfrac{7}{5}$

15. $\dfrac{2\frac{1}{2}}{3\frac{1}{2}} = \dfrac{\frac{5}{2}}{\frac{7}{2}} = \dfrac{5}{2} \cdot \dfrac{2}{7} = \dfrac{5}{7}$

17. $\dfrac{2\frac{2}{3}}{\frac{5}{3}} = \dfrac{\frac{8}{3}}{\frac{5}{3}} = \dfrac{8}{3} \cdot \dfrac{3}{5} = \dfrac{8}{5}$

19. $\dfrac{0.05}{0.15} = \dfrac{0.05 \times 100}{0.15 \times 100} = \dfrac{5}{15} = \dfrac{1}{3}$

21. $\dfrac{0.3}{3} = \dfrac{0.3 \times 10}{3 \times 10} = \dfrac{3}{30} = \dfrac{1}{10}$

23. $\dfrac{1.2}{10} = \dfrac{1.2 \times 10}{10 \times 10} = \dfrac{12}{100} = \dfrac{3}{25}$

25. a. $\dfrac{\text{shaded squares}}{\text{nonshaded squares}} = \dfrac{20}{40} = \dfrac{1}{2}$

   b. $\dfrac{\text{shaded squares}}{\text{total squares}} = \dfrac{20}{60} = \dfrac{1}{3}$

   c. $\dfrac{\text{nonshaded squares}}{\text{total squares}} = \dfrac{40}{60} = \dfrac{2}{3}$

Applying the Concepts

27. a. $\dfrac{\$650}{\$400} = \dfrac{650}{400} = \dfrac{13}{8}$

   b. $\dfrac{\$100}{\$400} = \dfrac{100}{400} = \dfrac{1}{4}$

   c. $\dfrac{\$150}{\$400} = \dfrac{150}{400} = \dfrac{3}{8}$

   d. $\dfrac{\$650}{\$150} = \dfrac{650}{150} = \dfrac{13}{3}$

29. a. $\dfrac{\text{Q1 revenue}}{\text{Q1 profit}} = \dfrac{\$6,000}{\$1,000} = \dfrac{6}{1}$

   b. $\dfrac{\text{Q2 revenue}}{\text{Q2 profit}} = \dfrac{\$7,500}{\$1,500} = \dfrac{5}{1}$

   c. $\dfrac{\text{Q3 revenue}}{\text{Q3 profit}} = \dfrac{\$8,400}{\$2,100} = \dfrac{4}{1}$

   d. $\dfrac{\text{Q4 revenue}}{\text{Q4 profit}} = \dfrac{\$10,500}{\$3,500} = \dfrac{3}{1}$

   e. $\dfrac{\text{annual revenue}}{\text{annual profit}} = \dfrac{\$32,400}{\$8,100} = \dfrac{4}{1}$

Calculator Problems

31. $\dfrac{2{,}314}{2{,}408} \approx 0.96$

33. $\dfrac{2{,}314}{4{,}722} \approx 0.49$

Review Problems

35. $\dfrac{90}{5} = \dfrac{90 \div 5}{5 \div 5} = \dfrac{18}{1} = 18$

37.
$$
\begin{array}{r}
62.5 \\
2{\overline{\smash{\big)}\,125.0}} \\
\underline{12\!\downarrow} \\
5 \\
\underline{4\!\downarrow} \\
10 \\
\underline{10} \\
0
\end{array}
$$

39.
$$
\begin{array}{r}
0.615 \\
2{\overline{\smash{\big)}\,1.230}} \\
\underline{12\!\downarrow} \\
3 \\
\underline{2\!\downarrow} \\
10 \\
\underline{10} \\
0
\end{array}
$$

41.
$$
0.25{\overline{\smash{\big)}\,46.0}} \;\to\; 25{\overline{\smash{\big)}\,4600}}
$$
$$
\begin{array}{r}
184 \\
25{\overline{\smash{\big)}\,4600}} \\
\underline{25\!\downarrow} \\
210 \\
\underline{200\!\downarrow} \\
100 \\
\underline{100} \\
0
\end{array}
$$

Improving Your Quantitative Literacy

43. IBM $\dfrac{P}{E} = \dfrac{\$146.05}{\$6.35} = \dfrac{14{,}605}{635} = \dfrac{23}{1} = 23$

AO $\dfrac{P}{E} = \dfrac{\$139.69}{\$0.61} = \dfrac{13{,}969}{61} = \dfrac{229}{1} = 229$

DIS $\dfrac{P}{E} = \dfrac{\$30.03}{\$0.91} = \dfrac{3{,}003}{91} = \dfrac{33}{1} = 33$

KM $\dfrac{P}{E} = \dfrac{\$15.64}{\$0.68} = \dfrac{1{,}564}{68} = \dfrac{23}{1} = 23$

GE $\dfrac{P}{E} = \dfrac{\$90.75}{\$2.75} = \dfrac{9{,}075}{275} = \dfrac{33}{1} = 33$

TOY $\dfrac{P}{E} = \dfrac{\$19.92}{\$1.66} = \dfrac{1{,}992}{166} = \dfrac{12}{1} = 12$

Undervalued stocks: IBM, KM, TOY

Section 4.2

1. $\dfrac{220 \text{ miles}}{4 \text{ hours}} = 55\dfrac{\text{miles}}{\text{hour}} = 55 \text{ miles per hour}$

3. $\dfrac{252 \text{ kilometers}}{3 \text{ hours}} = 84\dfrac{\text{kilometers}}{\text{hour}}$
$= 84 \text{ kilometers per hour}$

5. $\dfrac{3 \text{ gallons}}{15 \text{ seconds}} = 0.2\dfrac{\text{gallons}}{\text{second}}$
$= 0.2 \text{ gallons per second}$

7. $\dfrac{56 \text{ liters}}{4 \text{ minutes}} = 14\dfrac{\text{liters}}{\text{minute}}$
$= 14 \text{ liters per minute}$

9. $\dfrac{95 \text{ miles}}{5 \text{ gallons}} = 19\dfrac{\text{miles}}{\text{gallon}} = 19 \text{ miles per gallon}$

11. $\dfrac{325 \text{ miles}}{75 \text{ liters}} = 4\dfrac{1}{3}\dfrac{\text{miles}}{\text{liter}} = 4\dfrac{1}{3} \text{ miles per liter}$

13. $\dfrac{96¢}{6 \text{ ounces}} = 16¢ \text{ per ounce}$

15. $\dfrac{99¢}{20 \text{ ounces}} = 4.95¢ \text{ per ounce}$

17. Huggies:
$\dfrac{\$12.49}{36 \text{ diapers}} \approx \dfrac{\$0.347}{\text{diaper}} = 34.7¢ \text{ per diaper}$
Cruisers:
$\dfrac{\$11.99}{38 \text{ diapers}} \approx \dfrac{\$0.316}{\text{diaper}} = 31.6¢ \text{ per diaper}$
Cruisers are a better buy.

19. $\dfrac{675.4 \text{ miles}}{12\frac{1}{2} \text{ hours}} = \dfrac{675.4 \text{ miles}}{12.5 \text{ hours}}$
$\approx 54.03\dfrac{\text{miles}}{\text{hour}}$
$= 54.03 \text{ miles per hour}$

21. $\dfrac{128.4 \text{ miles}}{13.8 \text{ gallons}} \approx 9.3\dfrac{\text{miles}}{\text{gallon}}$
$= 9.3 \text{ miles per gallon}$

23. Daily rate of pay $= \dfrac{\$64}{\text{day}} = \$64 \text{ per day}$

25. Annual rate of pay
$= \dfrac{\$320}{\text{week}} \times \dfrac{50 \text{ weeks}}{\text{year}} = \dfrac{\$16,000}{\text{year}}$

Review Problems

27. The solution to $2 \cdot n = 12$ is $n = 6$, because $2 \cdot 6 = 12$.

29. The solution to $6 \cdot n = 24$ is $n = 4$, because $6 \cdot 4 = 24$.

31. The solution to $20 = 5 \cdot n$ is $n = 4$, because $20 = 5 \cdot 4$.

33. The solution to $650 = 10 \cdot n$ is $n = 65$, because $650 = 10 \cdot 65$.

Improving Your Quantitative Literacy

35. Old: $\dfrac{\$6.99}{100 \text{ oz}} = \dfrac{\$6.99}{3.12 \text{ quart}} \approx \dfrac{\$2.24}{\text{quart}}$

New: $\dfrac{\$5.75}{80 \text{ oz}} = \dfrac{\$5.75}{2.5 \text{ quart}} = \dfrac{\$2.30}{\text{quart}}$

The 100 ounce jug is the better buy.

Section 4.3

1. $\dfrac{3 \cdot 5 \cdot 5 \cdot 7}{3 \cdot 5} = \dfrac{\cancel{3} \cdot \cancel{5} \cdot 5 \cdot 7}{\cancel{3} \cdot \cancel{5}} = 5 \cdot 7 = 35$

3. $\dfrac{2 \cdot n \cdot 3 \cdot 3 \cdot 5}{n \cdot 5} = \dfrac{2 \cdot \cancel{n} \cdot 3 \cdot 3 \cdot \cancel{5}}{\cancel{n} \cdot \cancel{5}} = 2 \cdot 3 \cdot 3 = 18$

5. $\dfrac{2 \cdot 2 \cdot n \cdot 7 \cdot 11}{2 \cdot n \cdot 11} = \dfrac{\cancel{2} \cdot 2 \cdot \cancel{n} \cdot 7 \cdot \cancel{11}}{\cancel{2} \cdot \cancel{n} \cdot \cancel{11}} = 2 \cdot 7 = 14$

7. $\dfrac{9 \cdot n}{9} = \dfrac{\cancel{9} \cdot n}{\cancel{9}} = n$

9. $\dfrac{4 \cdot y}{4} = \dfrac{\cancel{4} \cdot y}{\cancel{4}} = y$

11. $4 \cdot n = 8$

$\dfrac{\cancel{4} \cdot n}{\cancel{4}} = \dfrac{8}{4}$

$n = 2$

13. $5 \cdot x = 35$

$\dfrac{\cancel{5} \cdot x}{\cancel{5}} = \dfrac{35}{5}$

$x = 7$

15. $3 \cdot y = 21$

$\dfrac{\cancel{3} \cdot y}{\cancel{3}} = \dfrac{21}{3}$

$y = 7$

17. $6 \cdot n = 48$

$\dfrac{\cancel{6} \cdot n}{\cancel{6}} = \dfrac{48}{6}$

$n = 8$

19. $5 \cdot a = 40$

$\dfrac{\cancel{5} \cdot a}{\cancel{5}} = \dfrac{40}{5}$

$a = 8$

21. $3 \cdot x = 6$

$\dfrac{\cancel{3} \cdot x}{\cancel{3}} = \dfrac{6}{3}$

$x = 2$

23. $2 \cdot y = 2$

$\dfrac{\cancel{2} \cdot y}{\cancel{2}} = \dfrac{2}{2}$

$y = 1$

25. $3 \cdot a = 18$

$\dfrac{\cancel{3} \cdot a}{\cancel{3}} = \dfrac{18}{3}$

$a = 6$

27. $5 \cdot n = 25$

$\dfrac{\cancel{5} \cdot n}{\cancel{5}} = \dfrac{25}{5}$

$n = 5$

29. $6 = 2 \cdot x$

$\dfrac{6}{2} = \dfrac{\cancel{2} \cdot x}{\cancel{2}}$

$3 = x$

31. $42 = 6 \cdot n$

$\dfrac{42}{6} = \dfrac{\cancel{6} \cdot n}{\cancel{6}}$

$7 = n$

33. $4 = 4 \cdot y$

$\dfrac{4}{4} = \dfrac{\cancel{4} \cdot y}{\cancel{4}}$

$1 = y$

35. $63 = 7 \cdot y$

$$\frac{63}{7} = \frac{\cancel{7} \cdot y}{\cancel{7}}$$

$$9 = y$$

37. $2 \cdot n = 7$

$$\frac{\cancel{2} \cdot n}{\cancel{2}} = \frac{7}{2}$$

$$n = \frac{7}{2} \text{ or } 3\frac{1}{2}$$

39. $6 \cdot x = 21$

$$\frac{\cancel{6} \cdot x}{\cancel{6}} = \frac{21}{6}$$

$$x = \frac{21}{6}$$

$$x = \frac{7}{2} \text{ or } 3\frac{1}{2}$$

41. $5 \cdot a = 12$

$$\frac{\cancel{5} \cdot a}{\cancel{5}} = \frac{12}{5}$$

$$a = \frac{12}{5} \text{ or } 2\frac{2}{5}$$

43. $4 = 7 \cdot y$

$$\frac{4}{7} = \frac{\cancel{7} \cdot y}{\cancel{7}}$$

$$\frac{4}{7} = y$$

45. $10 = 13 \cdot y$

$$\frac{10}{13} = \frac{\cancel{13} \cdot y}{\cancel{13}}$$

$$\frac{10}{13} = y$$

47. $12 \cdot x = 30$

$$\frac{\cancel{12} \cdot x}{\cancel{12}} = \frac{30}{12}$$

$$x = \frac{30}{12}$$

$$x = \frac{5}{2}$$

$$x = 2\frac{1}{2}$$

49. $21 = 14 \cdot n$

$$\frac{21}{14} = \frac{\cancel{14} \cdot n}{\cancel{14}}$$

$$\frac{21}{14} = n$$

$$\frac{3}{2} = n$$

$$1\frac{1}{2} = n$$

Calculator Problems

51. $24 \cdot x = 1{,}176$

$$\frac{\cancel{24} \cdot x}{\cancel{24}} = \frac{1{,}176}{24}$$

$$x = 49$$

53. $41 \cdot y = 7{,}831$

$$\frac{\cancel{41} \cdot y}{\cancel{41}} = \frac{7{,}831}{41}$$

$$y = 191$$

55. $436.6 = 37 \cdot n$

$$\frac{436.6}{37} = \frac{\cancel{37} \cdot n}{\cancel{37}}$$

$$11.8 = n$$

Review Problems

57. $$\begin{array}{r} 0.75 \\ 4\overline{)3.00} \\ \underline{28\downarrow} \\ 20 \\ \underline{20} \\ 0 \end{array} \qquad \frac{3}{4} = 0.75$$

59. $$5\frac{1}{2} = \frac{11}{2} \rightarrow \begin{array}{r} 5.5 \\ 2\overline{)11.0} \\ \underline{10\downarrow} \\ 10 \\ \underline{10} \\ 0 \end{array} \qquad 5\frac{1}{2} = \frac{11}{2} = 5.5$$

61. $$\begin{array}{r} 0.03 \\ 100\overline{)3.00} \\ \underline{100} \\ 0 \end{array} \qquad \frac{3}{100} = 0.03$$

63. $$0.34 = \frac{34}{100} = \frac{17}{50}$$

65. $$2.4 = 2\frac{4}{10} = 2\frac{2}{5}$$

67. $$1.75 = 1\frac{75}{100} = 1\frac{3}{4}$$

Section 4.4

1. Means: 3,5
Extremes: 1,15
$$\frac{1}{3} = \frac{5}{15}$$
$$1 \cdot 15 = 3 \cdot 5$$
$$15 = 15 \quad \text{True}$$

3. Means: 25,2
Extremes: 10,5
$$\frac{10}{25} = \frac{2}{5}$$
$$10 \cdot 5 = 25 \cdot 2$$
$$50 = 50 \quad \text{True}$$

5. Means: $\frac{1}{2}, 4$

Extremes: $\frac{1}{3}, 6$

$$\frac{\frac{1}{3}}{\frac{1}{2}} = \frac{4}{6}$$
$$\frac{1}{2} \cdot 4 = \frac{1}{3} \cdot 6$$
$$2 = 2 \quad \text{True}$$

7. Means: 5,1
Extremes: 0.5,10
$$\frac{0.5}{5} = \frac{1}{10}$$
$$0.5(10) = 5(1)$$
$$5 = 5 \quad \text{True}$$

9. $$\frac{2}{5} = \frac{4}{x}$$
$$2 \cdot x = 5 \cdot 4$$
$$2x = 20$$
$$x = 10$$

11. $$\frac{1}{y} = \frac{5}{12}$$
$$1 \cdot 12 = y \cdot 5$$
$$12 = 5y$$
$$\frac{12}{5} = y$$

13. $$\frac{x}{4} = \frac{3}{8}$$
$$x \cdot 8 = 4 \cdot 3$$
$$8x = 12$$
$$x = \frac{12}{8} = \frac{3}{2}$$

15. $$\frac{5}{9} = \frac{x}{2}$$
$$5 \cdot 2 = 9 \cdot x$$
$$10 = 9x$$
$$\frac{10}{9} = x$$

17. $$\frac{3}{7} = \frac{3}{x}$$
$$3 \cdot x = 7 \cdot 3$$
$$3x = 21$$
$$x = 7$$

19. $$\frac{25}{100} = \frac{x}{4}$$
$$25 \cdot 4 = 100 \cdot x$$
$$100 = 100x$$
$$1 = x$$

21.
$$\frac{\frac{1}{2}}{y} = \frac{\frac{1}{3}}{12}$$
$$\frac{1}{2} \cdot 12 = y \cdot \frac{1}{3}$$
$$6 = \frac{1}{3}y$$
$$18 = y$$

23.
$$\frac{n}{12} = \frac{\frac{1}{4}}{\frac{1}{2}}$$
$$n \cdot \frac{1}{2} = 12 \cdot \frac{1}{4}$$
$$\frac{1}{2}n = 3$$
$$n = 6$$

25.
$$\frac{10}{20} = \frac{20}{n}$$
$$10 \cdot n = 20 \cdot 20$$
$$10n = 400$$
$$n = 40$$

27.
$$\frac{x}{10} = \frac{10}{2}$$
$$x \cdot 2 = 10 \cdot 10$$
$$2x = 100$$
$$x = 50$$

29.
$$\frac{1}{100} = \frac{y}{50}$$
$$1 \cdot 50 = 100 \cdot y$$
$$50 = 100y$$
$$\frac{50}{100} = y$$
$$\frac{1}{2} = y$$

31.
$$\frac{0.4}{1.2} = \frac{1}{x}$$
$$0.4(x) = 1.2(1)$$
$$0.4x = 1.2$$
$$x = \frac{1.2}{0.4}$$
$$x = 3$$

33.
$$\frac{0.3}{0.18} = \frac{n}{0.6}$$
$$0.3(0.6) = 0.18(n)$$
$$0.18 = 0.18n$$
$$1 = n$$

35.
$$\frac{0.5}{x} = \frac{1.4}{0.7}$$
$$0.5(0.7) = x(1.4)$$
$$0.35 = 1.4x$$
$$\frac{0.35}{1.4} = x$$
$$0.25 = x$$
$$\frac{1}{4} = x$$

37.
$$\frac{168}{324} = \frac{56}{x}$$
$$168 \cdot x = 324 \cdot 56$$
$$168x = 18,144$$
$$x = 108$$

39.
$$\frac{429}{y} = \frac{858}{130}$$
$$429 \cdot 130 = y \cdot 858$$
$$55,770 = 858y$$
$$65 = y$$

41.

$$\frac{n}{39} = \frac{533}{507}$$

$$n \cdot 507 = 39 \cdot 533$$

$$507n = 20{,}787$$

$$n = 41$$

43.

$$\frac{756}{903} = \frac{x}{129}$$

$$756 \cdot 129 = 903x$$

$$97{,}524 = 903x$$

$$108 = x$$

Review Problems

45. 250.14
     ↑

    tens

47.
$$\begin{array}{r} 2.300 \\ 0.180 \\ +\ 24.036 \\ \hline 26.516 \end{array}$$

49.
$$\begin{array}{r} 3.18 \\ -\ 2.79 \\ \hline 0.39 \end{array}$$

Improving Your Quantitative Literacy

51. Answers will vary.

Section 4.5

1. Let x = distance traveled in 7 hours
$$\frac{235 \text{ miles}}{5 \text{ hours}} = \frac{x \text{ miles}}{7 \text{ hours}}$$
$$235 \cdot 7 = 5 \cdot x$$
$$1{,}645 = 5x$$
$$329 = x$$
She will travel 329 miles.

3. Let x = points scored in 20 games
$$\frac{162 \text{ points}}{9 \text{ games}} = \frac{x \text{ points}}{20 \text{ games}}$$
$$162 \cdot 20 = 9 \cdot x$$
$$3{,}240 = 9x$$
$$360 = x$$
At that rate, he will score 360 points.

5. Let x = pints of water
$$\frac{8 \text{ pts antifreeze}}{5 \text{ pts water}} = \frac{24 \text{ pts antifreeze}}{x \text{ pints water}}$$
$$8 \cdot x = 5 \cdot 24$$
$$8x = 120$$
$$x = 15$$
You need 15 pints of water to get the same concentration.

7. Let x = actual distance between the cities
$$\frac{1 \text{ in}}{95 \text{ miles}} = \frac{4.5 \text{ in}}{x \text{ miles}}$$
$$1(x) = 95(4.5)$$
$$x = 427.5$$
The distance between the cities is 427.5 miles.

9. Let x = number of marketable eggs
$$\frac{50 \text{ laid}}{45 \text{ marketable}} = \frac{1{,}000 \text{ laid}}{x \text{ marketable}}$$
$$50 \cdot x = 45 \cdot 1{,}000$$
$$50x = 45{,}000$$
$$x = 900$$
900 eggs will be marketable.

11. Let x = length of actual box car
$$\frac{1 \text{ in model}}{87 \text{ in actual}} = \frac{5 \text{ in model}}{x \text{ in actual}}$$
$$1 \cdot x = 87 \cdot 5$$
$$x = 435$$
The actual box car is 435 inches, which is 36.25 feet

13. Let x = cost for a week
$$\frac{21\cent}{1 \text{ mile}} = \frac{x \cent}{570 \text{ miles}}$$
$$21 \cdot 570 = 1 \cdot x$$
$$11{,}970 = x$$
It will cost him 11,970¢, or $119.70

15. Let x = grams of water in 10 ounces
$$\frac{6 \text{ oz serving}}{159 \text{ g of water}} = \frac{10 \text{ oz serving}}{x \text{ g of water}}$$
$$6 \cdot x = 159 \cdot 10$$
$$6x = 1{,}590$$
$$x = 265$$
There are 265 grams of water in a ten ounce serving.

17. Let x = liters of gas for 692 miles
$$\frac{378.9 \text{ miles}}{50 \text{ liters}} = \frac{692 \text{ miles}}{x \text{ liters}}$$
$$378.9(x) = 50(692)$$
$$378.9x = 34{,}600$$
$$x = 91.3 \text{ (to the nearest tenth)}$$
You would need 91.3 liters of gas.

19. Let x = number of people who voted
$$\frac{47 \text{ voters}}{100 \text{ registered}} = \frac{x \text{ voted}}{127{,}900 \text{ registered}}$$
$$47 \cdot 127{,}900 = 100 \cdot x$$
$$6{,}011{,}300 = 100x$$
$$60{,}113 = x$$
You would expect 60,113 people voted.

Review Problems

21.   2.7

$\underline{\times\ 0.5}$

   1.35

23.   3.18

$\underline{\times\ 1.2}$

    636

$\underline{3,180}$

  3.816

25. $\dfrac{2.8}{0.7} = \dfrac{2.8 \times 10}{0.7 \times 10} = \dfrac{28}{7} = 4$

27. $2.33\overline{)24} \rightarrow 15\overline{)2400}$

$$\begin{array}{r} 160 \\ \overline{)2400} \\ \underline{15\downarrow} \\ 90 \\ \underline{90\downarrow} \\ 00 \end{array}$$

Improving Your Quantitative Literacy

29.   Answers will vary.

Section 4.6

1. $\dfrac{6}{h} = \dfrac{4}{6}$

$36 = 4h$

$9 = h$

3. $\dfrac{y}{8} = \dfrac{21}{12}$

$12y = 168$

$y = 14$

5. $\dfrac{12}{16} = \dfrac{9}{x}$

$12x = 144$

$x = 12$

7. $\dfrac{5}{a} = \dfrac{3}{15}$

$3a = 75$

$a = 25$

9. $\dfrac{50}{40} = \dfrac{40}{y}$

$50y = 1{,}600$

$y = 32$

11. $\dfrac{BA}{AC} = \dfrac{ED}{DF}$

$\dfrac{4}{3} = \dfrac{ED}{6}$

$24 = 3 \cdot ED$

$8 = ED$

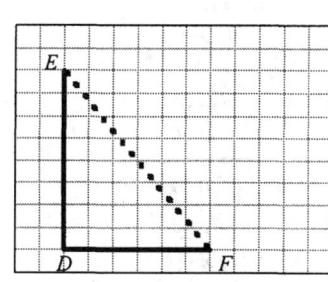

13. Let D be the midpoint of BC, and let H be the midpoint of EF.

$\dfrac{BD}{DA} = \dfrac{EH}{HG}$

$\dfrac{3}{6} = \dfrac{4}{HG}$

$3 \cdot HG = 24$

$HG = 8$

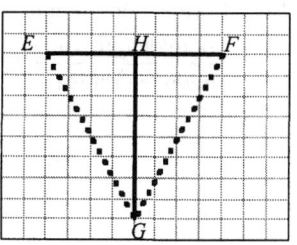

15. $\dfrac{AD}{DC} = \dfrac{EH}{HG}$

$\dfrac{3}{4} = \dfrac{EH}{8}$

$4 \cdot EH = 24$

$EH = 6$

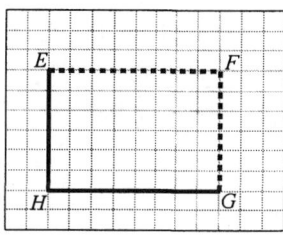

17. $\dfrac{AB}{AE} = \dfrac{FG}{FJ}$

$\dfrac{2}{4} = \dfrac{4}{FJ}$

$2 \cdot FJ = 16$

$FJ = 8$

Let L be the midpoint of AE, and let K be the midpoint of FJ.

$\dfrac{AL}{LC} = \dfrac{FK}{KH}$

$\dfrac{2}{4} = \dfrac{4}{KH}$

$2 \cdot KH = 16$

$KH = 8$

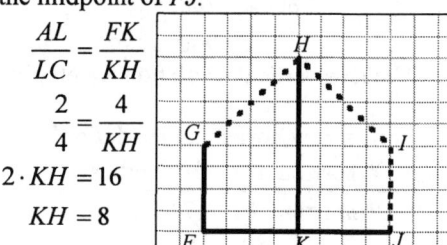

Applying the Concepts

19. Let x = body length of the Bass

$$\frac{\text{body length viola}}{\text{body length Bass}} = \frac{\text{total length viola}}{\text{total length Bass}}$$

$$\frac{15}{x} = \frac{24}{72}$$

$$24x = 1{,}080$$

$$x = 45$$

The body length of the Bass is 45 inches.

21. Let x = body length of the viola

$$\frac{\text{body length viola}}{\text{body length cello}} = \frac{\text{total length viola}}{\text{total length cello}}$$

$$\frac{x}{30} = \frac{26}{48}$$

$$48x = 780$$

$$x = 16.25$$

The body length of the viola is 16.25 inches.

23. Let x = new vertical resolution

$$\frac{\text{old horizontal}}{\text{new horizontal}} = \frac{\text{old vertical}}{\text{new vertical}}$$

$$\frac{800}{1{,}280} = \frac{600}{x}$$

$$800x = 768{,}000$$

$$x = 960$$

The new vertical resolution is 960 pixels.

25. Let x = new horizontal resolution

$$\frac{20'' \text{ horizontal}}{17'' \text{ horizontal}} = \frac{20'' \text{ vertical}}{17'' \text{ vertical}}$$

$$\frac{1{,}680}{1{,}050} = \frac{x}{900}$$

$$1{,}050x = 1{,}512{,}000$$

$$x = 1{,}440$$

The horizontal resolution of 1,440 pixels.

27. Let x = height of tree

$$\frac{\text{height of tree}}{\text{height of man}} = \frac{\text{tree shadow}}{\text{man shadow}}$$

$$\frac{x}{6} = \frac{38}{4}$$

$$4x = 228$$

$$x = 57$$

The tree is 57 feet tall.

29. Let x = height of the child

$$\frac{\text{height of tower}}{\text{height of child}} = \frac{\text{tower shadow}}{\text{child shadow}}$$

$$\frac{36}{x} = \frac{54}{6}$$

$$216 = 54x$$

$$4 = x$$

The child is 4 feet tall.

Review Problems

31.
$$
\begin{array}{r}
2.03 \\
11.958 \\
+\ 0.002 \\
\hline
13.990 \text{ or } 13.99
\end{array}
$$

33.
$$
\begin{array}{r}
65.002 \\
-\ 24.003 \\
\hline
40.999
\end{array}
$$

35.
$$
\begin{array}{r}
42.18 \\
\times\ 0.0025 \\
\hline
21090 \\
84360 \\
\hline
0.105450
\end{array}
$$
Drop the trailing 0: 0.10545

37.

$$21.05\overline{)378.9} \rightarrow 2105\overline{)37890}$$

$$
\begin{array}{r}
18 \\
2105\overline{)37890} \\
\underline{2105}\!\downarrow \\
16840 \\
\underline{16840} \\
0
\end{array}
$$

Extending the Concepts

39. a. $\dfrac{\text{shaded}}{\text{non-shaded}} = \dfrac{8}{16} = \dfrac{1}{2}$

b. Let $x=$ number of shaded rectangles in the larger rectangle.

$\dfrac{8 \text{ shaded}}{24 \text{ total}} = \dfrac{x \text{ shaded}}{54 \text{ total}}$

$24x = 432$

$x = 18$

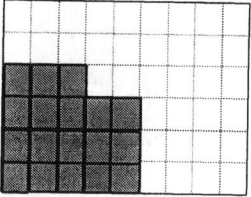

c. small rectangle: $\dfrac{\text{shaded}}{\text{total}} = \dfrac{8}{24} = \dfrac{1}{3}$

large rectangle: $\dfrac{\text{shaded}}{\text{total}} = \dfrac{18}{54} = \dfrac{1}{3}$

Chapter 4 Review

1. $\dfrac{9}{30} = \dfrac{3}{10}$

3. $\dfrac{\frac{3}{7}}{\frac{4}{7}} = \dfrac{3}{\cancel{7}} \cdot \dfrac{\cancel{7}}{4} = \dfrac{3}{4}$

5. $\dfrac{2\frac{1}{3}}{1\frac{2}{3}} = \dfrac{\frac{7}{3}}{\frac{5}{3}} = \dfrac{7}{\cancel{3}} \cdot \dfrac{\cancel{3}}{5} = \dfrac{7}{5}$

7. $\dfrac{0.6}{1.2} = \dfrac{0.6 \times 10}{1.2 \times 10} = \dfrac{6}{12} = \dfrac{1}{2}$

9. $\dfrac{\frac{1}{5}}{\frac{3}{5}} = \dfrac{1}{\cancel{5}} \cdot \dfrac{\cancel{5}}{3} = \dfrac{1}{3}$

11. $\dfrac{\text{taxes}}{\text{oil}} = \dfrac{18¢}{4¢} = \dfrac{9}{2}$

13. $\dfrac{\text{profits}}{\text{costs}} = \dfrac{4¢}{16¢} = \dfrac{1}{4}$

15. $\dfrac{285 \text{ miles}}{15 \text{ gallons}} = 19 \dfrac{\text{mi}}{\text{gal}}$

17. $\dfrac{138¢}{6 \text{ ounces}} = 23 \dfrac{¢}{\text{oz}} = 23¢ \text{ per ounce}$

19. $\dfrac{5}{7} = \dfrac{35}{x}$

$5 \cdot x = 7 \cdot 35$

$5x = 245$

$x = 49$

21. $\dfrac{\frac{1}{2}}{10} = \dfrac{y}{2}$

$\dfrac{1}{2} \cdot 2 = 10 \cdot y$

$1 = 10y$

$\dfrac{1}{10} = y$

23. Let x = mL of solution

$\dfrac{2{,}000 \text{ mL of solution}}{24 \text{ mL of drug}} = \dfrac{x \text{ mL of solution}}{18 \text{ mL of drug}}$

$2{,}000 \cdot 18 = 24 \cdot x$

$36{,}000 = 24x$

$1{,}500 = x$

$1{,}500$ mL of solution are required.

25. Let x = time to lose 20 pounds (in weeks)

$\dfrac{8 \text{ lbs}}{2 \text{ weeks}} = \dfrac{20 \text{ lbs}}{x \text{ weeks}}$

$8 \cdot x = 2 \cdot 20$

$8x = 40$

$x = 5$

It will take him 5 weeks.

27. $\dfrac{8}{6} = \dfrac{12}{x}$

$8x = 6 \cdot 12$

$8x = 72$

$x = 9$

Chapters 1-4 Cumulative Review

1.
$$\overset{1\ \ 1\ 1}{8,359}$$
$$401$$
$$+\ 1762$$
$$\overline{10,522}$$

3.
$$21\overline{)378} \ \overset{18}{}$$
$$\underline{21\downarrow}$$
$$168$$
$$\underline{168}$$
$$0$$

5.
$$31\overline{)15,689} \ \overset{506}{} \ \rightarrow \ \frac{15,689}{31} = 506\frac{3}{31}$$
$$\underline{155\downarrow\downarrow}$$
$$189$$
$$\underline{186}$$
$$3$$

7.
$$6 \cdot 2^3 - 1 = 6 \cdot 8 - 1$$
$$= 48 - 1$$
$$= 47$$

9.
$$56$$
$$+\ 18$$
$$\overline{74}$$

11.
$$(11 - 2) + (403 - 102) = 9 + 301$$
$$= 310$$

13.
$$83.60$$
$$-\ 12.12$$
$$\overline{71.48}$$

15.
$$6.8\overline{)30.6} \rightarrow 68\overline{)306.0} \ \overset{4.5}{}$$
$$\underline{272\downarrow}$$
$$340$$
$$\underline{340}$$
$$0$$

17.
$$5 \div \left(\frac{1}{4}\right)^2 = 5 \div \left(\frac{1}{4}\right)\left(\frac{1}{4}\right)$$
$$= 5 \div \frac{1}{16}$$
$$= 5 \cdot \frac{16}{1}$$
$$= 80$$

19.
$$5 \div \left(14 \div 1\frac{2}{3}\right) = 5 \div \left(\frac{14}{1} \div \frac{5}{3}\right)$$
$$= 5 \div \left(\frac{14}{1} \cdot \frac{3}{5}\right)$$
$$= 5 \div \frac{42}{5}$$
$$= \frac{5}{1} \cdot \frac{5}{42}$$
$$= \frac{25}{42}$$

21.
$$\frac{2}{5}(1.3) + \frac{1}{5}(2.1) = 0.4(1.3) + 0.2(2.1)$$
$$= 0.52 + 0.42$$
$$= 0.94$$

23. $11 - \dfrac{5}{6} \div \dfrac{2}{3} = 11 - \dfrac{5}{6} \cdot \dfrac{3}{2}$

$\qquad = 11 - \dfrac{5 \cdot \cancel{3}}{\cancel{3} \cdot 2 \cdot 2}$

$\qquad = 11 - \dfrac{5}{4}$

$\qquad = \dfrac{44}{4} - \dfrac{5}{4}$

$\qquad = \dfrac{39}{4}$ or $9\dfrac{3}{4}$

25. $\dfrac{9}{10} = \dfrac{18}{x}$

$9x = 180$

$\dfrac{\cancel{9}x}{\cancel{9}} = \dfrac{180}{9}$

$x = 20$

27. $P = $ sum of all sides

$\quad = 7.4 + 16.2 + 10.3$

$\quad = 33.9$ in

29. $\dfrac{660 \text{ words}}{30 \text{ minute}} = \dfrac{22 \text{ words}}{1 \text{ minute}} = 22$ words/min

31. 1/8 of the quotient of 160 and 10

$\quad \downarrow \downarrow \qquad\quad \downarrow$

$\quad \dfrac{1}{8} \quad \times \quad \dfrac{160}{10}$

$\dfrac{1}{8} \times \dfrac{160}{10} = \dfrac{\cancel{8} \cdot \cancel{10} \cdot 2}{\cancel{8} \cdot \cancel{10}} = 2$

33. The table is as follows:

Year	Finishers	To the nearest hundred
1975	339	300
1980	12,512	12,500
1985	15,881	15,900
1990	23,774	23,800
1995	26,754	26,800

35. $\dfrac{\$1.25}{6 \text{ cans}} \approx \dfrac{\$0.208}{\text{can}}$

$\dfrac{\$5.99}{24 \text{ cans}} \approx \dfrac{\$0.250}{\text{can}}$

The 6-pack of store brand is the better buy.

37. Let $x = $ wingspan of the actual plane

Use the given ratio: $\dfrac{1}{48} = \dfrac{\text{length of model}}{\text{length of plane}}$;

$\dfrac{1}{48} = \dfrac{10.25 \text{ in}}{x}$

$x = 492$ in

$x = 492 \ \cancel{\text{in}} \cdot \dfrac{1 \text{ ft}}{12 \ \cancel{\text{in}}} = 41$ ft

39. $\dfrac{12}{x} = \dfrac{8}{10}$

$120 = 8x$

$15 = x$

The length is $x = 15$ cm.

Chapter 4 Test

1. $\dfrac{24}{18} = \dfrac{4}{3}$

3. $\dfrac{5}{3\frac{1}{3}} = \dfrac{5}{\frac{10}{3}} = 5 \cdot \dfrac{3}{10} = \dfrac{15}{10} = \dfrac{3}{2}$

5. $\dfrac{\frac{3}{11}}{\frac{5}{11}} = \dfrac{3}{\cancel{11}} \cdot \dfrac{\cancel{11}}{5} = \dfrac{3}{5}$

7. $\dfrac{\text{phone payment}}{\text{food payment}} = \dfrac{\$60}{\$250} = \dfrac{6}{25}$

9. 16 oz can:

$\dfrac{\$2.59}{16 \text{ oz}} = \dfrac{259¢}{16 \text{ oz}} \approx 16.19¢ \text{ per ounce}$

12 oz can :

$\dfrac{\$1.89}{12 \text{ oz}} = \dfrac{189¢}{12 \text{ oz}} = 15.75¢ \text{ per ounce}$

The 12 ounce can is the better buy.

11. $\dfrac{1.8}{6} = \dfrac{2.4}{x}$

$1.8(x) = 6(2.4)$

$1.8x = 14.4$

$x = 8$

13. Let x = actual distance between the cities

$\dfrac{1 \text{ inches}}{60 \text{ miles}} = \dfrac{2\frac{1}{4} \text{ in}}{x \text{ miles}}$

$1(x) = 60(2.25)$

$x = 135$

The cities are 135 miles apart.

15. $\dfrac{20}{15} = \dfrac{h}{12}$

$240 = 15h$

$16 = h$

Chapter 4: A Glimpse of Algebra

1. Large rectangle: $l = x+4$, $w = x+2$
In the diagram, find the area of the 4 small rectangles:

	x	4
x	x^2	$4x$
2	$2x$	8

Area of large rectangle = sum of areas of small rectangles:
$$(x+4)(x+2) = x^2 + 4x + 2x + 8$$
$$= x^2 + 6x + 8$$

3. Large rectangle: $l = 2x+3$, $w = 3x+2$
In the diagram, find the area of the 4 small rectangles:

	$2x$	3
$3x$	$6x^2$	$9x$
2	$4x$	6

Area of large rectangle = sum of areas of small rectangles:
$$(2x+3)(3x+2) = 6x^2 + 9x + 4x + 6$$
$$= 6x^2 + 13x + 6$$

5. Large rectangle: $l = 7x+2$, $w = 3x+4$
In the diagram, find the area of the 4 small rectangles:

	$7x$	2
$3x$	$21x^2$	$6x$
4	$28x$	8

Area of large rectangle = sum of areas of small rectangles:
$$(7x+2)(3x+4) = 21x^2 + 6x + 28x + 8$$
$$= 21x^2 + 34x + 8$$

7.
$$(x+2)(x+3) = (x+2)x + (x+2)3$$
$$= x \cdot x + 2 \cdot x + x \cdot 3 + 2 \cdot 3$$
$$= x^2 + 2x + 3x + 6$$
$$= x^2 + 5x + 6$$

9.
$$(2x+3)(x+4) = 2x^2 + 3x + 8x + 12$$
$$= 2x^2 + 11x + 12$$

11.
$$(7x+3)(2x+5) = 14x^2 + 6x + 35x + 15$$
$$= 14x^2 + 41x + 15$$

13.
$$(3x+2)(3x+2) = 9x^2 + 6x + 6x + 4$$
$$= 9x^2 + 12x + 4$$

15.
$$(4a+5)(a+1) = 4a^2 + 5a + 4a + 5$$
$$= 4a^2 + 9a + 5$$

17.
$$(7y+8)(6y+9) = 42y^2 + 48y + 63y + 72$$
$$= 42y^2 + 111y + 72$$

19.
$$(4+6x)(2+3x) = 8 + 12x + 12x + 18x^2$$
$$= 8 + 24x + 18x^2$$

Chapter 5 Preview

1.
$$136.00$$
$$\underline{+\ 5.44}$$
$$141.44$$

3.
$$1{,}793{,}000$$
$$\underline{-\ 315{,}568}$$
$$1{,}477{,}432$$

5. $0.2 \times 100 = 20$

7.
$$0.15$$
$$\underline{\times\ 63}$$
$$45$$
$$\underline{900}$$
$$9.45$$

9. $3.62 \div 100 = \dfrac{3.62}{100} = 0.0362$

11.
$$600 \times 0.04 \times \frac{60}{360} = \frac{600}{1} \cdot \frac{4}{100} \cdot \frac{60}{360}$$
$$= \frac{\cancel{100} \cdot \cancel{6} \cdot 4 \cdot \cancel{60}}{1 \cdot \cancel{100} \cdot \cancel{60} \cdot \cancel{6}}$$
$$= 4$$

13. $\dfrac{45}{1{,}000} = \dfrac{9 \cdot \cancel{5}}{200 \cdot \cancel{5}} = \dfrac{9}{200}$

15.
$$\begin{array}{r} 0.375 \\ 8\overline{)3.000} \\ \underline{24\downarrow} \\ 60 \\ \underline{56\downarrow} \\ 40 \\ \underline{40} \\ 0 \end{array}$$
$\dfrac{3}{8} = 0.375$

17. $2\dfrac{1}{2} = \dfrac{5}{2} \rightarrow$
$$\begin{array}{r} 2.5 \\ 2\overline{)5.0} \\ \underline{4\downarrow} \\ 10 \\ \underline{10} \\ 0 \end{array}$$
$2\dfrac{1}{2} = 2.5$

19.
$$0.12 \cdot n = 1{,}836$$
$$\frac{0.\cancel{12} \cdot n}{0.\cancel{12}} = \frac{1{,}836}{0.12}$$
$$n = \frac{1{,}836}{0.12}$$
$$n = 15{,}300$$

Chapter 5 Pretest

1. $68\% = \dfrac{68}{100} = 0.68$

3. $21.5\% = \dfrac{21.5}{100} = 0.215$

5. $0.386 = 0.386 \times 100\% = 38.6\%$

7. $33\% = \dfrac{33}{100}$

9. $8.5\% = \dfrac{8.5}{100} = \dfrac{85}{1,000} = \dfrac{17}{200}$

11. $\dfrac{4}{5} = \dfrac{4 \cdot 20}{5 \cdot 20} = \dfrac{80}{100} = 80\%$

13. What number is 5% of 24?
    $n = 0.05 \cdot 24$
    $n = 1.2$

15. 12 is 24% of what number?
    $12 = 0.24 \cdot n$

    $\dfrac{12}{0.24} = \dfrac{\cancel{0.24} \cdot n}{\cancel{0.24}}$

    $n = \dfrac{12}{0.24}$

    $n = 50$

17. What is 6% of $150,000?
    $n = 0.06 \cdot \$150,000$
    $n = \$9,000$

19. $I = P \times R \times T$
    $\quad = \$2,000 \times 0.015 \times 1$
    $\quad = \$30$

Original Principal:	$2,000
+ Interest:	+30
Total amount after 1 year:	$2,030

Section 5.1

1. $20\% = \dfrac{20}{100}$

3. $60\% = \dfrac{60}{100}$

5. $24\% = \dfrac{24}{100}$

7. $65\% = \dfrac{65}{100}$

9. $23\% = \dfrac{23}{100} = 0.23$

11. $92\% = \dfrac{92}{100} = 0.92$

13. $9\% = \dfrac{9}{100} = 0.09$

15. $3.4\% = \dfrac{3.4}{100} = 0.034$

17. $6.34\% = \dfrac{6.34}{100} = 0.0634$

19. $0.9\% = \dfrac{0.9}{100} = 0.009$

21. Move decimal 2 places right, and attach the % symbol:
$0.23 = 23.0\% = 23\%$

23. Move decimal 2 places right, and attach the % symbol:
$0.92 = 92.0\% = 92\%$

25. Move decimal 2 places right, and attach the % symbol:
$0.45 = 45.0\% = 45\%$

27. Move decimal 2 places right, and attach the % symbol:
$0.03 = 3.0\% = 3\%$

29. Move decimal 2 places right, and attach the % symbol:
$0.6 = 60.0\% = 60\%$

31. Move decimal 2 places right, and attach the % symbol:
$0.8 = 80.0\% = 80\%$

33. Move decimal 2 places right, and attach the % symbol:
$0.27 = 27.0\% = 27\%$

35. Move decimal 2 places right, and attach the % symbol:
$1.23 = 123.0\% = 123\%$

37. $60\% = \dfrac{60}{100} = \dfrac{3}{5}$

39. $75\% = \dfrac{75}{100} = \dfrac{3}{4}$

41. $4\% = \dfrac{4}{100} = \dfrac{1}{25}$

43. $26.5\% = \dfrac{26.5}{100} = \dfrac{26.5 \times 10}{100 \times 10} = \dfrac{265}{1,000} = \dfrac{53}{200}$

45. $71.87\% = \dfrac{71.87}{100} = \dfrac{71.87 \times 100}{100 \times 100} = \dfrac{7,187}{10,000}$

47. $0.75\% = \dfrac{0.75}{100} = \dfrac{0.75 \times 100}{100 \times 100} = \dfrac{75}{10,000} = \dfrac{3}{400}$

49. $6\frac{1}{4}\% = \dfrac{6\frac{1}{4}}{100} = \dfrac{\frac{25}{4}}{100} = \dfrac{25}{4} \times \dfrac{1}{100} = \dfrac{25}{400} = \dfrac{1}{16}$

51. $33\frac{1}{3}\% = \dfrac{33\frac{1}{3}}{100} = \dfrac{\frac{100}{3}}{100} = \dfrac{100}{3} \times \dfrac{1}{100} = \dfrac{1}{3}$

53. *Change denominator to 100, then switch to percent:*

$$\frac{1}{2} = \frac{1 \times 50}{2 \times 50} = \frac{50}{100} = 50\%$$

55. *Change denominator to 100, then switch to percent:*

$$\frac{3}{4} = \frac{3 \times 25}{4 \times 25} = \frac{75}{100} = 75\%$$

57.
$$3\overline{)1.00}^{\,0.33}$$
$$\underline{9\downarrow}$$
$$10$$
$$\underline{9}$$
$$1$$

$\dfrac{1}{3} = 0.\overline{3} = 33\frac{1}{3}\%$

59. *Change denominator to 100, then switch to percent:*

$$\frac{4}{5} = \frac{4 \times 20}{5 \times 20} = \frac{80}{100} = 80\%$$

61.
$$8\overline{)7.000}^{\,0.875}$$
$$\underline{64\downarrow}$$
$$60$$
$$\underline{56\downarrow}$$
$$40$$
$$\underline{40}$$
$$0$$

$\dfrac{7}{8} = 0.875 = 87.5\%$

63. *Change denominator to 100, then switch to percent:*

$$\frac{7}{50} = \frac{7 \times 2}{50 \times 2} = \frac{14}{100} = 14\%$$

65. *Change denominator to 100, then switch to percent:*

$$3\frac{1}{4} = \frac{13}{4} = \frac{13 \times 25}{4 \times 25} = \frac{325}{100} = 325\%$$

67. *Change denominator to 100, then switch to percent:*

$$1\frac{1}{2} = \frac{3}{2} = \frac{3 \times 50}{2 \times 50} = \frac{150}{100} = 150\%$$

69.
$$43\overline{)21.0000}^{\,0.4883}$$
$$\underline{172\downarrow}$$
$$380$$
$$\underline{344\downarrow}$$
$$360$$
$$\underline{344}$$
$$160$$
$$\underline{129}$$
$$31$$

$\dfrac{21}{43} \approx 0.4883 \approx 48.8\%$

Applying the Concepts

71. $50\% = \dfrac{50}{100} = 0.50$

$75\% = \dfrac{75}{100} = 0.75$

73. a. $10\% = \dfrac{10}{100} = \dfrac{1}{10}$

$\qquad 4\% = \dfrac{4}{100} = \dfrac{1}{25}$

  b. $10\% = \dfrac{10}{100} = 0.1$

$\qquad 4\% = \dfrac{4}{100} = 0.04$

  c. $\dfrac{10\%}{4\%} = \dfrac{0.10}{0.04} = \dfrac{0.10 \times 10}{0.04 \times 10} = \dfrac{10}{4} = 2.5$

  Boys are $2\dfrac{1}{2}$ times more likely to have ADHD than girls.

75. *Change denominator to 100, then switch to percent:*

$\dfrac{1}{5} = \dfrac{1 \times 20}{5 \times 20} = \dfrac{20}{100} = 20\%$

77. Science and Mathematics:

$\dfrac{3}{25} = \dfrac{3 \cdot 4}{25 \times 4} = \dfrac{12}{100} = 12\%$

Agriculture:

$\dfrac{11}{50} = \dfrac{11 \times 2}{50 \times 2} = \dfrac{22}{100} = 22\%$

Architecture and Environmental Design:

$\dfrac{1}{10} = \dfrac{1 \times 10}{10 \times 10} = \dfrac{10}{100} = 10\%$

Business:

$\dfrac{3}{20} = \dfrac{3 \times 5}{20 \times 5} = \dfrac{15}{100} = 15\%$

Engineering:

$\dfrac{1}{4} = \dfrac{1 \times 25}{4 \times 25} = \dfrac{25}{100} = 25\%$

Liberal Arts:

$\dfrac{4}{25} = \dfrac{4 \times 4}{25 \times 4} = \dfrac{16}{100} = 16\%$

Calculator Problems

79. $\dfrac{29}{37} \approx 0.7838 = 78.4\%$

81. $\dfrac{6}{51} \approx 0.1176 = 11.8\%$

83. $\dfrac{236}{327} \approx 0.7217 = 72.2\%$

85. $\dfrac{1}{12} \approx 0.08333 = 8.3\%$

$\qquad \dfrac{1}{450} \approx 0.0022 = 0.2\%$

Review Problems

87.
$$
\begin{array}{r}
300 \\
\times\, 0.25 \\
\hline
1500 \\
6000 \\
\hline
75.00
\end{array}
$$

The trailing 0's can be dropped: 75

89.
$$
\begin{array}{r}
0.53 \\
\times\, 16 \\
\hline
318 \\
530 \\
\hline
8.48
\end{array}
$$

91. $0.05\overline{)25} \rightarrow 5\overline{)2500}$ $\quad \dfrac{25}{0.05} = 500$

$$
\begin{array}{r}
500 \\
5\overline{)2500} \\
25\downarrow \\
\hline
00
\end{array}
$$

93. $0.25\overline{)45} \rightarrow 25\overline{)4500}$ $\dfrac{45}{0.25} = 180$

$$25\!\downarrow$$
$$\overline{200}$$
$$200\!\downarrow$$
$$\overline{00}$$

(with quotient 180)

95. $3 \cdot n = 12$

$$\dfrac{3n}{3} = \dfrac{12}{3}$$

$$n = 4$$

Improving your Quantitative Literacy

97. a. $100\% - 55\% = 45\%$

b. $100\% - 37\% = 63\%$

c. Answers will vary.

Section 5.2

1. What number is 25% of 32?

$n = 0.25 \cdot 32$

$n = 8$

3. What number is 20% of 120?

$n = 0.20 \cdot 120$

$n = 24$

5. What number is 54% of 38?

$n = 0.54 \cdot 38$

$n = 20.52$

7. What number is 11% of 67?

$n = 0.11 \cdot 67$

$n = 7.37$

9. What percent of 24 is 12?

$n \cdot 24 = 12$

$\dfrac{n \cdot \cancel{24}}{\cancel{24}} = \dfrac{12}{24}$

$n = \dfrac{12}{24}$

$n = 0.5 = 50\%$

11. What percent of 50 is 5?

$n \cdot 50 = 5$

$\dfrac{n \cdot \cancel{50}}{\cancel{50}} = \dfrac{5}{50}$

$n = \dfrac{5}{50}$

$n = 0.1 = 10\%$

13. What percent of 36 is 9?

$n \cdot 36 = 9$

$\dfrac{n \cdot \cancel{36}}{\cancel{36}} = \dfrac{9}{36}$

$n = \dfrac{9}{36}$

$n = 0.25 = 25\%$

15. What percent of 8 is 6?

$n \cdot 8 = 6$

$\dfrac{n \cdot \cancel{8}}{\cancel{8}} = \dfrac{6}{8}$

$n = \dfrac{6}{8}$

$n = 0.75 = 2 = 75\%$

17. 32 is 50% of what number?

$32 = 0.50 \cdot n$

$\dfrac{32}{0.50} = \dfrac{\cancel{0.50} \cdot n}{\cancel{0.50}}$

$\dfrac{32}{0.50} = n$

$64 = n$

19. 10 is 20% of what number?

$10 = 0.20 \cdot n$

$\dfrac{10}{0.20} = \dfrac{\cancel{0.20} \cdot n}{\cancel{0.20}}$

$\dfrac{10}{0.20} = n$

$50 = n$

21. 37 is 4% of what number?

$37 = 0.04 \cdot n$

$\dfrac{37}{0.04} = \dfrac{\cancel{0.04} \cdot n}{\cancel{0.04}}$

$\dfrac{37}{0.04} = n$

$925 = n$

23. 8 is 2% of what number?

$8 = 0.02 \cdot n$

$\dfrac{8}{0.02} = \dfrac{\cancel{0.02} \cdot n}{\cancel{0.02}}$

$\dfrac{8}{0.02} = n$

$400 = n$

25. What is 6.4% of 87?

$$n = 0.064 \cdot 87$$

$$n = 5.568$$

27. 25% of what number is 30?

$$0.25 \cdot n = 30$$

$$\frac{\cancel{0.25} \cdot n}{\cancel{0.25}} = \frac{30}{0.25}$$

$$n = \frac{30}{0.25}$$

$$n = 120$$

29. 28% of 49 is what number?

$$0.28 \cdot 49 = n$$

$$13.72 = n$$

31. 27 is 120% of what number?

$$27 = 1.20 \cdot n$$

$$\frac{27}{1.20} = \frac{\cancel{1.20} \cdot n}{\cancel{1.20}}$$

$$\frac{27}{1.20} = n$$

$$22.5 = n$$

33. 65 is what percent of 130?

$$65 = n \cdot 130$$

$$\frac{65}{130} = \frac{n \cdot \cancel{130}}{\cancel{130}}$$

$$\frac{65}{130} = n$$

$$n = 0.5 = 50\%$$

35. What is 0.4% of 235,671?

$$n = 0.004 \cdot 235{,}671$$

$$n = 942.684$$

37. 4.89% of 2,000 is what number?

$$0.0489 \cdot 2{,}000 = n$$

$$97.8 = n$$

Applying the Concepts

39. Total calories: 210
Fat calories: 10
10 is what percent of 210?

$$10 = n \cdot 210$$

$$\frac{10}{210} = \frac{n \cdot \cancel{210}}{\cancel{210}}$$

$$\frac{10}{210} = n$$

$$n \approx 0.0476 = 4.8\% \text{ (to the nearest tenth)}$$

Because the fat calories are less than 30% of total calories, the food is considered healthy.

41. Total calories: 20
Fat calories: 10
10 is what percent of 20?

$$10 = n \cdot 20$$

$$\frac{10}{20} = \frac{n \cdot \cancel{20}}{\cancel{20}}$$

$$\frac{10}{20} = n$$

$$n \approx 0.5 = 50\%$$

Because the fat calories are more than 30% of total calories, the food not considered healthy.

Calculator Problems

43. What number is 62.5% of 398?

$$n = 0.625 \cdot 398$$

$$n = 248.75$$

45. What percent of 789 is 204?

$$n \cdot 789 = 204$$

$$\frac{n \cdot \cancel{789}}{\cancel{789}} = \frac{204}{789}$$

$$n = \frac{204}{789}$$

$$n \approx 0.2585 = 25.9\% \text{ (to the nearest tenth)}$$

47. 19 is 38.4% of what number?

$$19 = 0.384 \cdot n$$

$$\frac{19}{0.384} = \frac{\cancel{0.384} \cdot n}{\cancel{0.384}}$$

$$\frac{19}{0.384} = n$$

$$n = 49.5 \text{ (to the nearest tenth)}$$

Review Problems

49. $45 = 9 \cdot 5$

$\quad\quad = 3 \cdot 3 \cdot 5$

51. $105 = 5 \cdot 21$

$\quad\quad = 5 \cdot 3 \cdot 7$

$\quad\quad = 3 \cdot 5 \cdot 7$

53. $36 = 4 \cdot 9$

$\quad\quad = 2 \cdot 2 \cdot 3 \cdot 3$

55. $210 = 10 \cdot 21$

$\quad\quad = 2 \cdot 5 \cdot 3 \cdot 7$

$\quad\quad = 2 \cdot 3 \cdot 5 \cdot 7$

Extending the Concepts

57.

$$n \quad = \quad 0.25 \quad \cdot \quad 350$$
$$\downarrow \quad\quad \downarrow \quad \downarrow \quad\quad \downarrow \quad \downarrow$$

What number is 25% of 350?

59.

$$n \quad\quad \cdot \quad 24 \quad = \quad 16$$
$$\downarrow \quad\quad\quad \downarrow \quad \downarrow \quad \downarrow$$

What percent of 24 is 16?

61. $46 = 0.75 \quad \cdot \quad n$

$$\downarrow \;\downarrow \;\downarrow \quad\quad \downarrow \quad\quad\quad \downarrow$$

46 is 75% of what number?

63. Low estimate:
From eggs to hatchling:
What is 70% of 1,000?

$$n = 0.70(1000)$$

$$n = 700$$
From hatchling to fledgling:
What is 50% of 700?

$$n = 0.50(700)$$

$$n = 350$$
From hatchling to breeder:
What is 50% of 350?

$$n = 0.50(350)$$

$$n = 175$$

High estimate:
From eggs to hatchling:
What is 80% of 1,000?

$$n = 0.80(1,000)$$

$$n = 800$$
From hatchling to fledgling:
What is 70% of 800?

$$n = 0.70(800)$$

$$n = 560$$
From hatchling to breeder:
What is 50% of 350?

$$n = 0.50(560)$$

$$n = 280$$

Fewer than 175 to 280 gulls.

$$\frac{--}{75} = \frac{-\,\diagup}{\cancel{75}}$$

$$\frac{63}{75} = n$$

$$n = 0.84 = 84\%$$

5. What is 75% of 60 mL?
$$n = 0.75 \cdot 60$$

$$n = 45 \text{ mL}$$

7. What is 65% of 28 acres?
$$n = 0.65 \cdot 28$$

$$n = 18.2 \text{ acres}$$
18.2 acres are available for farming.
$$28 - 18.2 = 9.8$$
9.8 acres are not available for farming.

9. 1,440 is 48% of what number?
$$1,440 = 0.48 \cdot n$$

$$\frac{1,440}{0.48} = \frac{\cancel{0.48} \cdot n}{\cancel{0.48}}$$

$$\frac{1,440}{0.48} = n$$

$$n = 3,000 \text{ total students}$$

11. 240 is 60% of what number?
$$240 = 0.6 \cdot n$$

$$\frac{240}{0.6} = \frac{\cancel{0.6} \cdot n}{\cancel{0.6}}$$

$$\frac{240}{0.6} = n$$

$$n = 400 \text{ total students}$$

13. What is 52% of 3,200 students?
$$n = 0.52 \cdot 3,200$$

$$n = 1,664 \text{ female students}$$

15. 10,000 is what percent of 32,000?
$$10,000 = n \cdot 32,000$$

$$\frac{10,000}{32,000} = \frac{n \cdot \cancel{32,000}}{\cancel{32,000}}$$

$$\frac{10,000}{32,000} = n$$

$$n = 0.3125 = 31.25\%$$

Calculator Problems

17. 3,972 is what percent of 7,892?
$$3,972 = n \cdot 7,892$$

$$\frac{3,972}{7,892} = \frac{n \cdot \cancel{7,892}}{\cancel{7,892}}$$

$$\frac{3,972}{7,892} = n$$

$$n = 0.503 \approx 50\%$$

19. a. $119,648 + 6,742 + 506 + 77 = 126,973$

b. 126,973 is 3% of what number?
$$126,973 = 0.03 \cdot n$$
$$\frac{126,973}{0.03} = \frac{\cancel{0.03} \cdot n}{\cancel{0.03}}$$
$$\frac{126,973}{0.03} = n$$
$$n \approx 4,232,433 \text{ children born}$$

c. 119,648 is what percent of 126,973?
$$119,648 = n \cdot 126,973$$
$$\frac{119,648}{126,973} = \frac{n \cdot \cancel{126,973}}{\cancel{126,973}}$$
$$\frac{119,648}{126,973} = n$$
$$n \approx 0.9423 \approx 94.23\%$$

Review Problems

21. $\dfrac{1}{2} \cdot \dfrac{2}{5} = \dfrac{1 \cdot \cancel{2}}{\cancel{2} \cdot 5} = \dfrac{1}{5}$

23. $\dfrac{3}{4} \cdot \dfrac{5}{9} = \dfrac{\cancel{3} \cdot 5}{4 \cdot 3 \cdot \cancel{3}} = \dfrac{5}{12}$

25. $2 \cdot \dfrac{3}{8} = \dfrac{\cancel{2} \cdot 3}{\cancel{2} \cdot 4} = \dfrac{3}{4}$

27. $1\dfrac{1}{4} \cdot \dfrac{8}{15} = \dfrac{5}{4} \cdot \dfrac{8}{15} = \dfrac{\cancel{5} \cdot 2 \cdot \cancel{4}}{\cancel{4} \cdot 3 \cdot \cancel{5}} = \dfrac{2}{3}$

Extending the Concepts

29. 55 is what percent of 163?
$$55 = n \cdot 163$$
$$\frac{55}{163} = \frac{n \cdot \cancel{163}}{\cancel{163}}$$
$$\frac{55}{163} = n$$
$$n \approx 0.3374 \approx 33.7\%$$

31. Note: Batting $0.333 = 33.3\%$
What is 33.3% of 132 at bats?
$$n = 0.333 \cdot 132$$
$$n = 44 \text{ hits (to the nearest whole)}$$

33. It is easier to work this problem using cumulative batting average versus the average for the next 50 at bats.

From the opening discussion on batting averages in the text, Suzuki had 72 hits in 200 at bats for an average of 0.360.

After the next 50 at bats, he would have a cumulative total of $200 + 50 = 250$ at bats.

Find the number of hits needed to have 0.360 (36.0%) out of 250 at bats:
What is 36% of 250 at bats?
$$n = 0.36 \cdot 250$$
$$n = 90 \text{ hits}$$
He needs 90 total hits. Because he already has 72 hits, he must have at least $90 - 72 = 18$ hits to maintain his average.

Section 5.4

1. What is 7% of $750?

$n = 0.07 \cdot 750$

$n = 52.5$

The sales tax is $52.50

3. What is 6% of $45?

$n = 0.06 \cdot 45$

$n = 2.7$

The sales tax is $2.70

Total price = purchase price + sales tax

$$= \$45 + \$2.70$$

$$= \$47.70$$

5. 4% of what number is $6?

$0.04 \cdot n = 6$

$$\frac{0.04 n}{0.04} = \frac{6}{0.04}$$

$n = \$150$

The purchase price is $150.

Total price = purchase price + sales tax

$$= \$150 + \$6$$

$$= \$156$$

7. $22.50 is what percent of $450?

$22.50 = n \cdot 450$

$$\frac{22.50}{450} = \frac{n \cdot 450}{450}$$

$$n = \frac{22.5}{450}$$

$n = 0.05 = 5\%$

9. What is 3% of $94,000?

$n = 0.03 \cdot 94{,}000$

$n = 2{,}820$

The commission is $2,820

11. 12% of what number is $24?

$0.12 \cdot n = 24$

$$\frac{0.12 n}{0.12} = \frac{24}{0.12}$$

$n = 200$

The purchase price is $200.

13. $112 is what percent of $800?

$112 = n \cdot 800$

$$\frac{112}{800} = \frac{n \cdot 800}{800}$$

$$n = \frac{112}{800}$$

$n = 0.14 = 14\%$

Calculator Problems

15. What is 5.5% of $216.95?

$n = 0.055 \cdot 216.95$

$n = 11.9323$

The sales tax is $11.93

17. $10.33 is what percent of $229.50?

$10.33 = n \cdot 229.50$

$$\frac{10.33}{229.50} = \frac{n \cdot 229.50}{229.50}$$

$$n = \frac{10.33}{229.50}$$

$n = 0.04501 = 4.5\%$

19. 13% of what number is $519.35?

$0.13 \cdot n = 519.35$

$$\frac{0.13 n}{0.13} = \frac{519.35}{0.13}$$

$n = \$3{,}995$

The car sold for $3,995

Review Problems

21. $\dfrac{1}{3} \div \dfrac{2}{3} = \dfrac{1}{3} \cdot \dfrac{3}{2} = \dfrac{1 \cdot 3}{3 \cdot 2} = \dfrac{1}{2}$

23. $2 \div \dfrac{3}{4} = 2 \cdot \dfrac{4}{3} = \dfrac{2 \cdot 4}{3} = \dfrac{8}{3}$ or $2\dfrac{2}{3}$

25. $\dfrac{3}{8} \div \dfrac{1}{4} = \dfrac{3}{8} \cdot \dfrac{4}{1} = \dfrac{3 \cdot \cancel{4}}{2 \cdot \cancel{4} \cdot 1} = \dfrac{3}{2}$ or $1\dfrac{1}{2}$

27. $2\dfrac{1}{4} \div \dfrac{1}{2} = \dfrac{9}{4} \cdot \dfrac{2}{1} = \dfrac{9 \cdot \cancel{2}}{\cancel{2} \cdot 2 \cdot 1} = \dfrac{9}{2} = 4\dfrac{1}{2}$

Extending the Concepts: Luxury Taxes

29. Sales tax:
What is 6% of $53,000?
$n = 0.06 \cdot 53,000$

$n = \$3,180$
To find luxury tax:

Purchase Price:	$53,000
$-\$30,000:$	$-30,000$
Amount over $30,000:	$23,000

Luxury tax:
What is 10% of $23,000?
$n = 0.1 \cdot 23,000$

$n = \$2,300$

31. You would have saved the luxury tax of
$2,300

33.
Sticker Price:	$31,500
$-$ Re duced Price:	$- \; 29,900$
Sticker savings:	$1,600

Luxury Tax savings:

Purchase Price:	$31,500
$-\$30,000:$	$-30,0000$
Amount over $30,000:	$1,500

Luxury tax:
What is 10% of $1,500
$n = 0.1 \cdot 1,500$

$n = \$150$

You save $1,600 on the sticker price and
$150 in luxury tax. If you live in a state
with a 6% sales tax, you saved an additional
$96, as shown below:

Sales Tax savings:
What is 6% of $1,600
$n = 0.06 \cdot 1,600$

$n = \$96$

Section 5.5

1. Increase:
    What is 7% of $23,000?
    $n = 0.07 \cdot 23,000$

    $n = \$1,610$

Old Salary:	$23,000
+ Increase:	+1,610
New Salary:	$24,610

3. Increase:
    What is 17% of $3,000?
    $n = 0.17 \cdot 3,000$

    $n = \$510$

Present tuition:	$3,000
+ Increase:	+510
New tuition:	$3,510

5. Decrease:
    What is 20% of $16,500?
    $n = 0.2 \cdot 16,500$

    $n = \$3,300$

New price:	$16,500
− Decrease:	−3,300
Value after 1 year:	$13,200

7. 350 is what percent of 3,500?
    $350 = n \cdot 3,500$

    $$\frac{350}{3,500} = \frac{n \cdot 3,500}{3,500}$$

    $$n = \frac{350}{3,500}$$

    $n = 0.1 = 10\%$

9. Decrease = $25 − $20 = $5
    Percent Decrease:
    $5 is what percent of $25?
    $5 = n \cdot 25$

    $$\frac{5}{25} = \frac{n \cdot 25}{25}$$

    $$n = \frac{5}{25}$$

    $n = 0.2 = 20\%$ decrease

11. a. Increase = $39.51 − $25.97 = $13.54
        Percent Increase:
        $13.54 is what percent of $25.97?
        $13,54 = n \cdot 25.97$

        $$\frac{13.54}{25.97} = \frac{n \cdot 25.97}{25.97}$$

        $$n = \frac{13.54}{25.97}$$

        $n \approx 0.5214 \approx 52.1\%$ increase

    b. Increase = $39.51 − $36.39 = $3.12
        Percent Increase:
        $3.12 is what percent of $36.39?
        $3.12 = n \cdot 36.39$

        $$\frac{3.12}{36.39} = \frac{n \cdot 36.39}{36.39}$$

        $$n = \frac{3.12}{36.39}$$

        $n \approx 0.0857 \approx 8.6\%$ increase

13. Discount:
    What is 15% of $300?
    $n = 0.15 \cdot 300$

    $n = \$45$
    The discount is $45.

Original price:	$300
− Discount:	− 45
Discount Price:	$255

15. Discount:
What is 20% of $450?
$n = 0.20 \cdot 450$

$n = \$90$
Original price: $450
   − Discount: − 90
   ───────────────────
   Sale Price: $360
Sales tax:
What is 6% of $360?
$n = 0.06 \cdot 360$

$n = \$21.60$
   Sale price: $360.00
   + Sales tax: + 21.60
   ───────────────────
   Total bill: $381.60

Calculator Problems

17. Increase:
What is 6.5% of $43,752?
$n = 0.065 \cdot 43,752$

$n = \$2,843.88$
   Old Salary: $43,752.00
   + Increase: + 2,843.88
   ───────────────────
   New Salary: $46,595.88

19. Area of smallest field:
$$A = lw = (100 \text{ yd})(55 \text{ yd}) = 5,500 \text{ yd}^2$$
Area of Rose Bowl:
$$A = lw = (116 \text{ yd})(72 \text{ yd}) = 8,352 \text{ yd}^2$$
Area of largest field:
$$A = lw = (120 \text{ yd})(75 \text{ yd}) = 9,000 \text{ yd}^2$$

a.   Increase $= 8,352 \text{ yd}^2 - 5,500 \text{ yd}^2$
$$= 2,852 \text{ yd}^2$$
Percent Increase:
$2,852 \text{ yd}^2$ is what percent of $5,500 \text{ yd}^2$?
$2,852 = n \cdot 5,500$

$$\frac{2,852}{5,500} = \frac{n \cdot \cancel{5,500}}{\cancel{5,500}}$$

$$n = \frac{2,852}{5,500}$$

$$n \approx 0.5185 \approx 51.9\% \text{ increase}$$

b.   Increase $= 9,000 \text{ yd}^2 - 8,352 \text{ yd}^2$
$$= 648 \text{ yd}^2$$
Percent Increase:
648 yd^2 is what percent of $8,352 \text{ yd}^2$?
$648 = n \cdot 8,352$

$$\frac{648}{8,352} = \frac{n \cdot \cancel{8,352}}{\cancel{8,352}}$$

$$n = \frac{648}{8,352}$$

$$n \approx 0.0776 \approx 7.8\% \text{ increase}$$

Review Problems

21. $\dfrac{1}{3} + \dfrac{2}{3} = \dfrac{1+2}{3} = \dfrac{3}{3} = 1$

23. $\dfrac{1}{2} + \dfrac{1}{4} = \dfrac{1 \cdot 2}{2 \cdot 2} + \dfrac{1}{4} = \dfrac{2}{4} + \dfrac{1}{4} = \dfrac{2+1}{4} = \dfrac{3}{4}$

25. $\dfrac{3}{4} + \dfrac{2}{3} = \dfrac{3 \cdot 3}{4 \cdot 3} + \dfrac{2 \cdot 4}{3 \cdot 4} = \dfrac{9}{12} + \dfrac{8}{12} = \dfrac{17}{12} = 1\dfrac{5}{12}$

27.

$$2\frac{1}{2}$$

$$+\,3\frac{1}{2}$$

$$5\frac{2}{2} = 5 + 1 = 6$$

Improving Your Quantitative Literacy

29.   a.   Increase = $5{,}458 - 517 = 4{,}941$

Percent Increase:

4,941 is what percent of 517?

$$4{,}941 = n \cdot 517$$

$$\frac{4{,}941}{517} = \frac{n \cdot \cancel{517}}{\cancel{517}}$$

$$n = \frac{4{,}941}{517}$$

$$n \approx 9.55705 \approx 956\% \text{ increase}$$

b.   Decrease = $\$1{,}533.85 - \$359.00$

$$= \$1{,}174.85$$

Percent decrease:

\$1,174.85 is what percent of \$1,533.85?

$$1{,}174.85 = n \cdot 1{,}533.85$$

$$\frac{1{,}174.85}{1{,}533.85} = \frac{n \cdot \cancel{1{,}533.85}}{\cancel{1{,}533.85}}$$

$$n = \frac{1{,}174.85}{1{,}533.85}$$

$$n \approx 0.76594 \approx 76.59\% \text{ decrease}$$

c.   No. A 100% decrease in price would mean the camcorders are free.

Section 5.6

1. $I = P \times R \times T$

$\quad = \$2,000 \times 0.08 \times 1$

$\quad = \$160$

Original Principal:	\$2,000
+ Interest:	+160
Total amount after 1 year:	\$2,160

3. $I = P \times R \times T$

$\quad = \$9,500 \times 0.07 \times 1$

$\quad = \$665$

5. $I = P \times R \times T$

$\quad = \$8,000 \times 0.07 \times 1$

$\quad = \$560$

Amount borrowed:	\$8,000
+ Interest:	+560
Total amount to pay back:	\$8,560

7. $I = P \times R \times T$

$\quad = \$2,000 \times 0.08 \times 1$

$\quad = \$160$

Amount borrowed:	\$2,000
+ Interest:	+160
Total amount to pay back:	\$2,160

9. $I = P \times R \times T$

$\quad = \$600 \times 0.05 \times \dfrac{60}{360}$

$\quad = \$600 \times 0.05 \times \dfrac{1}{6}$

$\quad = \$5$

11. $I = P \times R \times T$

$\quad = \$800 \times 0.05 \times \dfrac{120}{360}$

$\quad = \$800 \times 0.05 \times \dfrac{1}{3}$

$\quad = \$13.33 \ (\text{to nearest cent})$

Original Principal:	\$800.00
+ Interest:	+ 13.33
Total amount to withdraw:	\$813.33

13. Interest after 1st year:

$I = P \times R \times T$

$\quad = \$5,000 \times 0.06 \times 1$

$\quad = \$300$

Original Principal:	\$5,000
+ Interest:	+300
Total amount after 1 year:	\$5,300

Interest after 2nd year:

$I = P \times R \times T$

$\quad = \$5,300 \times 0.06 \times 1$

$\quad = \$318$

Amount at start of 2nd year:	\$5,300
+ Interest:	+318
Total amount after 2 years:	\$5,618

15. 1st Quarter:
$$I = P \times R \times T$$
$$= \$8{,}000 \times 0.05 \times \frac{1}{4}$$
$$= \$100$$

Original principal: $8,000
+ Interest: +100
New principal: $8,100

2nd quarter:
$$I = P \times R \times T$$
$$= \$8{,}100 \times 0.05 \times \frac{1}{4}$$
$$= \$101.25$$

Original principal: $8,100.00
+ Interest: +101.25
New principal: $8,201.25

3rd Quarter:
$$I = P \times R \times T$$
$$= \$8{,}201.25 \times 0.05 \times \frac{1}{4}$$
$$= \$102.52$$

Original principal: $8,201.25
+ Interest: +102.52
New principal: $8,303.77

4th Quarter:
$$I = P \times R \times T$$
$$= \$8{,}303.77 \times 0.05 \times \frac{1}{4}$$
$$= \$103.80$$

Original principal: $8,303.77
+ Interest: +103.80
New principal: $8,407.57

Calculator Problems

17.
$$I = P \times R \times T$$
$$= \$917.26 \times 0.0625 \times 1$$
$$= \$57.33$$

Original Principal: $917.26
+ Interest: + 57.33
Total amount after 1 year: $974.59

19a.
$$A = P\left(1 + \frac{r}{n}\right)^{n \cdot t}$$
$$= 10{,}000\left(1 + \frac{0.06}{4}\right)^{4 \cdot 5}$$
$$= 10{,}000(1 + 0.015)^{20}$$
$$= 10{,}000(1.015)^{20}$$
$$= 10{,}000(1.346855)$$
$$= \$13{,}468.55$$

19b.
$$A = P\left(1 + \frac{r}{n}\right)^{n \cdot t}$$
$$= 10{,}000\left(1 + \frac{0.06}{12}\right)^{12 \cdot 5}$$
$$= 10{,}000(1 + 0.005)^{60}$$
$$= 10{,}000(1.005)^{60}$$
$$= 10{,}000(1.3488501)$$
$$= \$13{,}488.50$$

19c.
$$A = P\left(1 + \frac{r}{n}\right)^{n \cdot t}$$
$$= 10{,}000\left(1 + \frac{0.05}{4}\right)^{4 \cdot 5}$$
$$= 10{,}000(1 + 0.0125)^{20}$$
$$= 10{,}000(1.0125)^{20}$$
$$= 10{,}000(1.282037232)$$
$$= \$12{,}820.37$$

19d. (continued)

$$A = P\left(1 + \frac{r}{n}\right)^{n \cdot t}$$

$$= 10,000\left(1 + \frac{0.05}{12}\right)^{12 \cdot 5}$$

$$= 10,000(1 + 0.004166667)^{60}$$

$$= 10,000(1.00416666667)^{60}$$

$$= 10,000(1.283358679)$$

$$= \$12,833.59$$

Review Problems

21. $\dfrac{3}{4} - \dfrac{1}{4} = \dfrac{3-1}{4} = \dfrac{2}{4} = \dfrac{1}{2}$

23. $\dfrac{5}{8} - \dfrac{1}{4} = \dfrac{5}{8} - \dfrac{1 \cdot 2}{4 \cdot 2} = \dfrac{5}{8} - \dfrac{2}{8} = \dfrac{5-2}{8} = \dfrac{3}{8}$

25. $\dfrac{1}{3} - \dfrac{1}{4} = \dfrac{1 \cdot 4}{3 \cdot 4} - \dfrac{1 \cdot 3}{4 \cdot 3} = \dfrac{4}{12} - \dfrac{3}{12} = \dfrac{4-3}{12} = \dfrac{1}{12}$

27. $3\dfrac{1}{4}$
$\underline{-\ 2\phantom{\dfrac{1}{4}}}$
$1\dfrac{1}{4}$

Extending the Concepts

29. Star Wars to The Empire Strikes Back:
Increase $= 18 - 11 = 7$
Percent Increase:
7 million is what percent of 11 million?
$7 = n \cdot 11$

$$\frac{7}{11} = \frac{n \cdot \cancel{11}}{\cancel{11}}$$

$$n = \frac{7}{11}$$

$$n = 0.636 = 63.6\% \ (\text{to nearest tenth})$$

The Empire Strikes Back to Return of the Jedi:
Increase $= 32.5 - 18 = 14.5$
Percent Increase:
14.5 million is what percent of 18 million?
$14.5 = n \cdot 18$

$$\frac{14.5}{18} = \frac{n \cdot \cancel{18}}{\cancel{18}}$$

$$n = \frac{14.5}{18}$$

$$n = 0.8055 = 80.6\% \ (\text{to nearest tenth})$$

Return of the Jedi to The Phantom Menace:
Increase $= 115 - 32.5 = 82.5$
Percent Increase:
82.5 million is what percent of 32.5 million?
$82.5 = n \cdot 32.5$

$$\frac{82.5}{32.5} = \frac{n \cdot \cancel{32.5}}{\cancel{32.5}}$$

$$n = \frac{82.5}{32.5}$$

$$n = 2.5384 = 253.8\% \ (\text{to nearest tenth})$$

Section 5.7

1. Note: each section of pie is worth 5%
    Number of A's:
    $$\frac{5}{20} = 0.25 = 25\% = 5 \text{ sections of the circle}$$
    Number of B's:
    $$\frac{8}{20} = 0.40 = 40\% = 8 \text{ sections of the circle}$$
    Number of C's:
    $$\frac{7}{20} = 0.35 = 35\% = 7 \text{ sections}$$

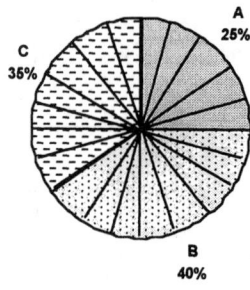

3. Note: each section of pie is worth 5%
    Kitchen:
    $$\frac{400}{2,400} \approx 0.17 = 17\% = \text{just over 3 sections}$$
    Dining Room:
    $$\frac{310}{2,400} \approx 0.13 = 13\% = \text{just over 2 sections}$$
    Bedrooms:
    $$\frac{890}{2,400} \approx 0.37 = 37\% = \text{just over 5 sections}$$
    Living Room:
    $$\frac{600}{2,400} = 0.25 = 25\% = 5 \text{ sections}$$
    Bathrooms:
    $$\frac{200}{2,400} = 0.08 = 8\% = \text{just over 1 section}$$

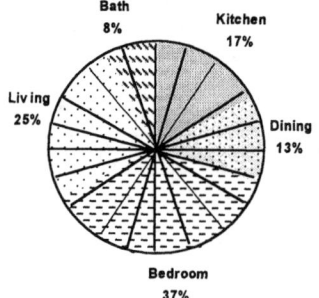

Review Problems

5. $$8 \cdot \frac{1}{3} = \frac{8}{1} \cdot \frac{1}{3} = \frac{8 \cdot 1}{1 \cdot 3} = \frac{8}{3} = 2\frac{2}{3}$$

7. $$25 \cdot \frac{1}{1,000} = \frac{25}{1} \cdot \frac{1}{1,000} = \frac{\cancel{25} \cdot 1}{1 \cdot \cancel{25} \cdot 40} = \frac{1}{40}$$

9. $36.5 \cdot \dfrac{1}{100} \cdot 10 = 36\dfrac{1}{2} \cdot \dfrac{1}{100} \cdot 10$

$$= \dfrac{73}{2} \cdot \dfrac{1}{100} \cdot \dfrac{10}{1}$$

$$= \dfrac{73 \cdot 1 \cdot \cancel{10}}{2 \cdot 10 \cdot \cancel{10}}$$

$$= \dfrac{73}{20} \text{ or } 3\dfrac{13}{20}$$

11. $248 \cdot \dfrac{1}{10} \cdot \dfrac{1}{10} = \dfrac{248}{10 \cdot 10} = \dfrac{248}{100} = \dfrac{62}{25} = 2\dfrac{12}{25}$

13. $48 \cdot \dfrac{1}{12} \cdot \dfrac{1}{3} = \dfrac{48 \cdot 1 \cdot 1}{12 \cdot 3} = \dfrac{\cancel{12} \cdot 4}{\cancel{12} \cdot 3} = \dfrac{4}{3} = 1\dfrac{1}{3}$

Chapter 5 Review

1. $35\% = \dfrac{35}{100} = 0.35$

3. $5\% = \dfrac{5}{100} = 0.05$

5. Move the decimal point 2 places right and attach the % sign:
$0.95 = 95.0\% = 95\%$

7. Move the decimal point 2 places right and attach the % sign:
$0.495 = 49.5\%$

9. $75\% = \dfrac{75}{100} = \dfrac{3}{4}$

11. $145\% = \dfrac{145}{100} = \dfrac{29}{20} = 1\dfrac{9}{20}$

13. $\dfrac{3}{10} = \dfrac{3\cdot10}{10\cdot10} = \dfrac{30}{100} = 30\%$

15.
$$3\overline{)2.000}$$
$$\begin{array}{r} 0.666 \\ \hline 2.000 \\ 18\!\downarrow \\ \hline 20 \\ 18\!\downarrow \\ \hline 20 \\ 18 \\ \hline 2 \end{array}$$
$\dfrac{2}{3} = 0.666... = 66\dfrac{2}{3}\%$

17. What number is 60% of 28?
$n = 0.60\cdot28$
$n = 16.8$

19. 24 is 30% of what number?
$24 = 0.30\cdot n$
$\dfrac{24}{0.30} = \dfrac{\cancel{0.30}\cdot n}{\cancel{0.30}}$
$\dfrac{24}{0.30} = n$
$80 = n$

21. a. What is 4.9% of 21,000 people?
$n = 0.049\cdot21{,}000$
$n = 1029$ people

  b. What is 10.1% of 21,000 people?
$n = 0.101\cdot21{,}000$
$n = 2{,}121$ people

  c. What is 74.1% of 280 million people?
$n = 0.741\cdot280{,}000{,}000$
$n = 207{,}480{,}000$ people

23. Discount:
What is 25% of $600?
$n = 0.25\cdot600$
$n = \$150$

Original price:	$600
− Discount:	−150
Sale Price:	$450

Sales tax:
What is 6% of $450?
$n = 0.06\cdot450$
$n = \$27$

Sale price:	$450
+ Sales tax:	+ 27
Total price:	$477

25. Increase:
What is 16% of $1.25?
$$n = 0.16 \cdot 1.25$$
$$n = \$0.20$$

Old price:	$1.25
+ increase:	+ 0.20
New price	$1.45

27. Increase = $906 − $583 = $323
Percent Increase:
$323 is what percent of $583?
$$\frac{323}{583} = \frac{n \cdot 583}{583}$$
$$n = \frac{323}{583} = 0.554 = 55.4\%$$
(to the nearest tenth of a percent)

29. $I = P \times R \times T$
$$= \$1,800 \times 0.07 \times \frac{120}{360}$$
$$= \$1,800 \times 0.07 \times \frac{1}{3}$$
$$= \$42$$

31. Note: each section of pie is worth 5%
Amount of used space:
$$\frac{150 \text{ MB}}{250 \text{ MB}} = 0.60 = 60\% = 12 \text{ sections of the}$$
circle
Amount of free space:
$$\frac{100 \text{ MB}}{250 \text{ MB}} = 0.40 = 40\% = 8 \text{ sections of the}$$
circle

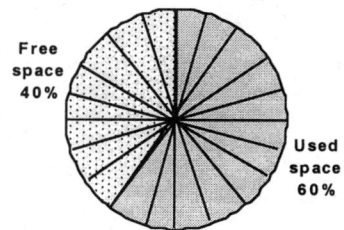

Chapters 1-5 Cumulative Review

1. $\begin{array}{r} 6,801 \\ 539 \\ +\ 374 \\ \hline 7,714 \end{array}$

3. $\begin{array}{r} 867 \\ \times\ 52 \\ \hline 1,734 \\ 43,350 \\ \hline 45,084 \end{array}$

5. $4.73\overline{)156.09} \rightarrow 473\overline{)15,609}$ with quotient 33
    $\begin{array}{r} 1419\downarrow \\ \hline 1419 \\ 1419 \\ \hline 0 \end{array}$

7. $\left(\dfrac{5}{6}\right)^3 = \dfrac{5}{6}\cdot\dfrac{5}{6}\cdot\dfrac{5}{6} = \dfrac{5\cdot5\cdot5}{6\cdot6\cdot6} = \dfrac{125}{216}$

9. $\begin{array}{r} 5.000 \\ -\ 3.678 \\ \hline 1.322 \end{array}$

11. $\dfrac{7}{15}\cdot\dfrac{5}{14} = \dfrac{\cancel{7}\cdot\cancel{5}}{3\cdot\cancel{5}\cdot2\cdot\cancel{7}} = \dfrac{1}{6}$

13. $\left.\begin{array}{l} 8 = 2\cdot2\cdot2 \\ 12 = 2\cdot2\cdot3\cdot3 \end{array}\right\}$ LCD $= 2\cdot2\cdot2\cdot3 = 24$

    $\dfrac{3}{8}+\dfrac{7}{12} = \dfrac{3\cdot3}{8\cdot3}+\dfrac{7\cdot2}{12\cdot2} = \dfrac{9}{24}+\dfrac{14}{24} = \dfrac{23}{24}$

15. $9\cdot4\dfrac{2}{3} = \dfrac{9}{1}\cdot\dfrac{14}{3} = \dfrac{\cancel{3}\cdot3\cdot14}{1\cdot\cancel{3}} = 42$

17. $\dfrac{1\frac{1}{2}}{\frac{1}{4}} = \dfrac{\frac{3}{2}}{\frac{1}{4}} = \dfrac{3}{2}\cdot\dfrac{4}{1} = \dfrac{3\cdot\cancel{2}\cdot2}{\cancel{2}\cdot1} = 6$

19. $\dfrac{3}{12} = \dfrac{\cancel{3}\cdot1}{\cancel{3}\cdot4} = \dfrac{1}{4}$

21. $1\text{ yd}^2 = 1\ \cancel{\text{yd}^2}\times\dfrac{9\ \cancel{\text{ft}^2}}{1\ \cancel{\text{yd}^2}}\times\dfrac{144\text{ in}^2}{1\ \cancel{\text{ft}^2}}$

    $= \dfrac{1\cdot9\cdot144}{1\cdot1}\text{ in}^2$

    $= 1,296\text{ in}^2$

    $\dfrac{1}{2}\text{ yd}^2 = \dfrac{1}{2}\ \cancel{\text{yd}^2}\times\dfrac{1,296\text{ in}^2}{1\ \cancel{\text{yd}^2}}$

    $= \dfrac{1,296}{2}\text{ in}^2$

    $= 648\text{ in}^2$

23. $46\% = \dfrac{46}{100} = \dfrac{23}{50}$

25. $3\cdot5^2 + 2\cdot4^2 - 5\cdot2^3 = 3\cdot25+2\cdot16-5\cdot8$
    $= 75+32-40$
    $= 67$

27. 55 is what percent of 275?
    $55 = n\cdot275$
    $\dfrac{55}{275} = \dfrac{n\cdot\cancel{275}}{\cancel{275}}$
    $n = \dfrac{55}{275} = 0.2 = 20\%$

29. $\dfrac{\$2.79}{6\text{ cans}} = \dfrac{\$0.465}{1\text{ can}} \approx 47\cancel{c}\text{ per serving}$

31. When $F = 212°$:

$$C = \frac{5(F-32)}{9}$$

$$= \frac{5(212-32)}{9}$$

$$= \frac{5(180)}{9}$$

$$= \frac{900}{9}$$

$$= 100$$

212°F is equivalent to 100°C.

33. Increase: $\$1,800 - \$1,600 = 200$

Percent increase:

$200 is what percent of $1,600?

$$200 = n \cdot 1,600$$

$$\frac{200}{1,600} = \frac{n \cdot 1,600}{1,600}$$

$$n = \frac{200}{1,600} = 0.125 = 12.5\%$$

She had a 12.5% increase in her monthly salary.

35. The product of 6 and 8 is $6 \cdot 8 = 48$

The sum of 6 and 8 is $6 + 8 = 14$

The difference between the product and the sum is $48 - 14 = 34$

37.

8.5 in

8.5 in

$$P = 4s \qquad A = s^2$$

$$= 4(8.5 \text{ in}) \qquad = (8.5 \text{ in})^2$$

$$= 34 \text{ in} \qquad = 72.25 \text{ in}^2$$

39. $\text{Hourly pay} = \dfrac{\$56 \text{ total wages}}{8 \text{ hours}}$

$$= \$7 \text{ per hour}$$

Chapter 5 Test

1. $18\% = \dfrac{18}{100} = 0.18$

3. $0.5\% = \dfrac{0.5}{100} = 0.005$

5. Move decimal point 2 places right and attach % sign:
$0.7 = 70.0\% = 70\%$

7. $65\% = \dfrac{65}{100} = \dfrac{13}{20}$

9. $3.5\% = \dfrac{3.5}{100} = \dfrac{3.5 \times 10}{100 \times 10} = \dfrac{35}{1,000} = \dfrac{7}{200}$

11.
$$\begin{array}{r} 0.375 \\ 8\overline{)3.000} \\ \underline{24\downarrow} \\ 60 \\ \underline{56\downarrow} \\ 40 \\ \underline{40} \\ 0 \end{array}$$
$\dfrac{3}{8} = 0.375 = 37.5\%$

13. What number is 75% of 60?
$n = 0.75 \cdot 60$
$n = 45$

15. 16 is 20% of what number?
$16 = 0.20 \cdot n$
$\dfrac{16}{0.20} = \dfrac{\cancel{0.20} \cdot n}{\cancel{0.20}}$
$\dfrac{16}{0.20} = n$
$80 = n$

17. What is 8% of $12,000?
$n = 0.08 \cdot 12,000$
$n = \$960$

19. Discount:
What is 25% of $280?
$n = 0.25 \cdot 280$
$n = \$70$

Original price: $280
$\underline{-\text{ Discount:} \quad -70}$
Sale Price: $210

Sales tax:
What is 5% of $210?
$n = 0.05 \cdot 210$
$n = \$10.50$

Sale price: $210.00
$\underline{+\text{ Sales tax:} \quad +\ 10.50}$
Total price: $220.50

21. At the end of 1st year:
$I = P \times R \times T$
$= \$12,000 \times 0.1 \times 1$
$= \$1,200$

Original Principal: $12,000
$\underline{+ \text{ Interest:} \quad +1,200}$
Total amount after 1 year: $13,200

At the end of 2nd year:
$I = P \times R \times T$
$= \$13,200 \times 0.1 \times 1$
$= \$1,320$

Interest from 1st year: $1,200
$\underline{+ \text{ Interest from 2nd year:} \quad +1,320}$
Total interest: $2,520

Chapter 5: A Glimpse of Algebra

1. Let $x = 2$:
$$6x + 2x - 7 = 6(2) + 2(2) - 7$$
$$= 12 + 4 - 7$$
$$= 9$$

3. Let $x = 10$:
$$4x + 6x + 8x = 4(10) + 6(10) + 8(10)$$
$$= 40 + 60 + 80$$
$$= 180$$

5. Let $a = 5$:
$$\frac{4a + 20}{5a - 20} = \frac{4(5) + 20}{5(5) - 20}$$
$$= \frac{20 + 20}{25 - 20}$$
$$= \frac{40}{5}$$
$$= 8$$

7. Let $a = 3$:
$$\frac{2a + 3a + 1}{4a + 5a + 3} = \frac{2(3) + 3(3) + 1}{4(3) + 5(3) + 3}$$
$$= \frac{6 + 9 + 1}{12 + 15 + 3}$$
$$= \frac{16}{30}$$
$$= \frac{8}{15}$$

9. Let $x = 2$:
$$x^2 + 5x + 6 = (2)^2 + 5(2) + 6$$
$$= 4 + 10 + 6$$
$$= 20$$

11. Let $x = 1$:
$$x^2 + 10x + 25 = (1)^2 + 10(1) + 25$$
$$= 1 + 10 + 25$$
$$= 36$$

13. Let $x = 5$ and $y = 2$:
$$\frac{3x + y}{3x - y} = \frac{3(5) + 2}{3(5) - 2}$$
$$= \frac{15 + 2}{15 - 2}$$
$$= \frac{17}{13}$$
$$= 1\frac{4}{13}$$

15. Let $x = 5$ and $y = 4$:
$$\frac{4x + 6y}{6x + 4y} = \frac{4(5) + 6(4)}{6(5) + 4(4)}$$
$$= \frac{20 + 24}{30 + 16}$$
$$= \frac{44}{46}$$
$$= \frac{22}{23}$$

17. Let $x = 4$:
$$(3x + 2)(3x - 2) = (3 \cdot 4 + 2)(3 \cdot 4 - 2)$$
$$= (12 + 2)(12 - 2)$$
$$= (14)(10)$$
$$= 140$$

19. Let $x = 1$:
$$(2x + 3)^2 = (2 \cdot 1 + 3)^2$$
$$= (2 + 3)^2$$
$$= 5^2$$
$$= 25$$

21. Let $x = 2$:

$$\frac{x^3 + 1}{x + 1} = \frac{2^3 + 1}{2 + 1}$$

$$= \frac{8 + 1}{2 + 1}$$

$$= \frac{9}{3}$$

$$= 3$$

23. Let $x = 3$:

$$\frac{x^3 - 8}{x^2 + 2x + 4} = \frac{(3)^3 - 8}{3^2 + 2(3) + 4}$$

$$= \frac{27 - 8}{9 + 6 + 4}$$

$$= \frac{19}{19}$$

$$= 1$$

25. Let $x = 5$:

$$\frac{x^4 - 16}{x^2 + 4} = \frac{5^4 - 16}{5^2 + 4}$$

$$= \frac{625 - 16}{25 + 4}$$

$$= \frac{609}{29}$$

$$= 21$$

Chapter 6 Preview

1. $\dfrac{12}{30} = \dfrac{\cancel{6}\cdot 2}{\cancel{6}\cdot 5} = \dfrac{2}{5}$

3.
$$\begin{array}{r} 12 \\ \times\ 16 \\ \hline 72 \\ 120 \\ \hline 192 \end{array}$$

5.
$$\begin{array}{r} 43,560 \\ \times\ 75 \\ \hline 217,800 \\ 3,049,200 \\ \hline 3,267,000 \end{array}$$

7.
$$\begin{array}{r} 2.49 \\ \times\ 3.75 \\ \hline 1,245 \\ 17,430 \\ 74,700 \\ \hline 9.3375 \end{array}$$

9. $8 \times \dfrac{1}{3} = \dfrac{8}{1}\cdot\dfrac{1}{3} = \dfrac{8\cdot 1}{1\cdot 3} = \dfrac{8}{3} = 2\dfrac{2}{3}$

11. $\dfrac{1800}{4} = \dfrac{450}{1} = 450$

13. $\dfrac{80.5}{1.61} = \dfrac{80.5\times 100}{1.61\times 100} = \dfrac{8,050}{161} = \dfrac{50}{1} = 50$

15. $\dfrac{1,100\times 60\times 60}{5,280} = \dfrac{3,960000}{5,280} = 750$

17. $\dfrac{2\cdot 1,000}{16.39} = \dfrac{2,000}{16.39} \approx 122$

19.
$$\begin{array}{r} 0.75 \\ 16\overline{)12.00} \\ \underline{112}\downarrow \\ 80 \\ \underline{80} \\ 0 \end{array}$$

Chapter 6 Pretest

1. $8 \text{ ft} = 8 \cancel{\text{ft}} \times \dfrac{12 \text{ in}}{1 \cancel{\text{ft}}}$

$\quad\quad = 8 \times 12 \text{ in}$

$\quad\quad = 96 \text{ in}$

3. $32 \text{ m} = 32 \cancel{\text{m}} \times \dfrac{100 \text{ cm}}{1 \cancel{\text{m}}}$

$\quad\quad = 32 \times 100 \text{ cm}$

$\quad\quad = 3,200 \text{ cm}$

5. $30 \text{ yd}^2 = 30 \cancel{\text{yd}^2} \times \dfrac{9 \text{ ft}^2}{1 \cancel{\text{yd}^2}}$

$\quad\quad = 30 \times 9 \text{ ft}^2$

$\quad\quad = 270 \text{ ft}^2$

7. $3,840 \text{ acres} = 3,840 \cancel{\text{acres}} \times \dfrac{1 \text{ mi}^2}{640 \cancel{\text{acres}}}$

$\quad\quad = \dfrac{3,840}{640} \text{mi}^2$

$\quad\quad = 6 \text{ mi}^2$

9. $3 \text{ gal} = 3 \cancel{\text{gal}} \times \dfrac{4 \text{ qt}}{1 \cancel{\text{gal}}}$

$\quad\quad = 3 \times 4 \text{ qt}$

$\quad\quad = 12 \text{ qt}$

11. $251 \text{ mL} = 251 \cancel{\text{mL}} \times \dfrac{1 \text{ L}}{1,000 \cancel{\text{mL}}}$

$\quad\quad = \dfrac{251}{1,000} \text{L}$

$\quad\quad = 0.251 \text{ L}$

13. $2,142 \text{ mg} = 2,142 \cancel{\text{mg}} \times \dfrac{1 \text{ g}}{1,000 \cancel{\text{mg}}}$

$\quad\quad = \dfrac{2,142}{1,000} \text{ g}$

$\quad\quad = 2.142 \text{ g}$

15. $3 \text{ gal} = 3 \cancel{\text{gal}} \times \dfrac{3.79 \text{ L}}{1 \cancel{\text{gal}}}$

$\quad\quad = 3 \times 3.79 \text{ L}$

$\quad\quad = 11.37 \text{ L}$

17. 45 miles per hour

$= 45 \dfrac{\cancel{\text{mi}}}{\cancel{\text{hr}}} \times \dfrac{5,280 \text{ ft}}{1 \cancel{\text{mi}}} \times \dfrac{1 \cancel{\text{hr}}}{60 \cancel{\text{min}}} \times \dfrac{1 \cancel{\text{min}}}{60 \text{ sec}}$

$= \dfrac{45 \times 5,280}{60 \times 60} \dfrac{\text{ft}}{\text{sec}}$

$= 66 \dfrac{\text{ft}}{\text{sec}}$

19. $5 \text{ gal} = 5 \cancel{\text{gal}} \times \dfrac{128 \cancel{\text{fl oz}}}{1 \cancel{\text{gal}}} \times \dfrac{1 \text{ glass}}{8 \cancel{\text{fl oz}}}$

$\quad\quad = \dfrac{5 \times 128}{8} \text{ glasses}$

$\quad\quad = 80 \text{ glasses}$

Section 6.1

1. $5 \text{ ft} = 5 \text{ ft} \times \dfrac{12 \text{ in}}{1 \text{ ft}}$

$\qquad = 5 \times 12 \text{ in}$

$\qquad = 60 \text{ in}$

3. $10 \text{ ft} = 10 \text{ ft} \times \dfrac{12 \text{ in}}{1 \text{ ft}}$

$\qquad = 10 \times 12 \text{ in}$

$\qquad = 120 \text{ in}$

5. $2 \text{ yd} = 2 \text{ yd} \times \dfrac{3 \text{ ft}}{1 \text{ yd}}$

$\qquad = 2 \times 3 \text{ ft}$

$\qquad = 6 \text{ ft}$

7. $4.5 \text{ yd} = 4.5 \text{ yd} \times \dfrac{3 \text{ ft}}{1 \text{ yd}} \times \dfrac{12 \text{ in}}{1 \text{ ft}}$

$\qquad = 4.5 \times 3 \times 12 \text{ in}$

$\qquad = 162 \text{ in}$

9. $27 \text{ in} = 27 \text{ in} \times \dfrac{1 \text{ ft}}{12 \text{ in}}$

$\qquad = \dfrac{27}{12} \text{ ft}$

$\qquad = 2\dfrac{1}{4} \text{ ft}$

11. $19 \text{ ft} = 19 \text{ ft} \times \dfrac{1 \text{ yd}}{3 \text{ ft}}$

$\qquad = \dfrac{19}{3} \text{ yd}$

$\qquad = 6\dfrac{1}{3} \text{ yd}$

13. $48 \text{ in} = 48 \text{ in} \times \dfrac{1 \text{ ft}}{12 \text{ in}} \times \dfrac{1 \text{ yd}}{3 \text{ ft}}$

$\qquad = 48 \times \dfrac{1}{12} \times \dfrac{1}{3} \text{ yd}$

$\qquad = 4 \times \dfrac{1}{3} \text{ yd}$

$\qquad = \dfrac{4}{3} \text{ yd}$

$\qquad = 1\dfrac{1}{3} \text{ yd}$

15. $18 \text{ m} = 18 \text{ m} \times \dfrac{100 \text{ cm}}{1 \text{ m}}$

$\qquad = 18 \times 100 \text{ cm}$

$\qquad = 1,800 \text{ cm}$

17. $4.8 \text{ km} = 4.8 \text{ km} \times \dfrac{1,000 \text{ m}}{1 \text{ km}}$

$\qquad = 4.8 \times 1,000 \text{ m}$

$\qquad = 4,800 \text{ m}$

19. $5 \text{ dm} = 5 \text{ dm} \times \dfrac{1 \text{ m}}{10 \text{ dm}} \times \dfrac{100 \text{ cm}}{1 \text{ m}}$

$\qquad = \dfrac{5 \times 100}{10} \text{ cm}$

$\qquad = 50 \text{ cm}$

21. $248 \text{ m} = 248 \text{ m} \times \dfrac{1 \text{ km}}{1,000 \text{ m}}$

$\qquad = \dfrac{248}{1,000} \text{ km}$

$\qquad = 0.248 \text{ km}$

23. $67 \text{ cm} = 67 \text{ cm} \times \dfrac{1 \text{ m}}{100 \text{ cm}} \times \dfrac{1,000 \text{ mm}}{1 \text{ m}}$

$\qquad = \dfrac{67 \times 1,000}{100} \text{ mm}$

$\qquad = 670 \text{ mm}$

25. $3,498 \text{ cm} = 3,498 \cancel{\text{cm}} \times \dfrac{1 \text{ m}}{100 \cancel{\text{cm}}}$

$= \dfrac{3,498}{100} \text{m}$

$= 34.98 \text{ m}$

27. $63.4 \text{ cm} = 63.4 \cancel{\text{cm}} \times \dfrac{1 \cancel{\text{m}}}{100 \cancel{\text{cm}}} \times \dfrac{10 \text{ dm}}{1 \cancel{\text{m}}}$

$= \dfrac{63.4 \times 10}{100} \text{dm}$

$= 6.34 \text{ dm}$

Applying the Concepts

29. $60 \text{ ft} = 60 \cancel{\text{ft}} \times \dfrac{1 \text{ yd}}{3 \cancel{\text{ft}}}$

$= \dfrac{60}{3} \text{ yd}$

$= 20 \text{ yd}$

31. First convert 6 feet to inches:

$6 \text{ ft} = 6 \cancel{\text{ft}} \times \dfrac{12 \text{ in}}{1 \cancel{\text{ft}}}$

$= 6 \times 12 \text{ in}$

$= 72 \text{ in}$

6 feet 8 inches = 72 in + 8 in = 80 in

33. $2.44 \text{ m} = 2.44 \cancel{\text{m}} \times \dfrac{100 \text{ cm}}{1 \cancel{\text{m}}}$

$= 2.44 \times 100 \text{ cm}$

$= 244 \text{ cm}$

35. $6.5 \text{ cm} = 6.5 \cancel{\text{cm}} \times \dfrac{1 \cancel{\text{m}}}{100 \cancel{\text{cm}}} \times \dfrac{1,000 \text{ mm}}{1 \cancel{\text{m}}}$

$= \dfrac{6.5 \times 1,000}{100} \text{mm}$

$= 65 \text{ mm}$

37. $37 \text{ mi} = 37 \cancel{\text{mi}} \times \dfrac{80 \text{ chains}}{1 \cancel{\text{mi}}}$

$= 37 \times 80 \text{ chains}$

$= 2,960 \text{ chains}$

39. 12 cm

$= 12 \cancel{\text{cm}} \times \dfrac{1 \cancel{\text{m}}}{100 \cancel{\text{cm}}} \times \dfrac{1,000 \cancel{\text{mm}}}{1 \cancel{\text{m}}} \times \dfrac{1,000 \mu m}{1 \cancel{\text{mm}}}$

$= \dfrac{12 \times 1,000 \times 1,000}{100} \mu m$

$= 120,000 \ \mu m$

41. 12 furlongs

$= 12 \ \cancel{\text{furlongs}} \times \dfrac{220 \text{ yds}}{1 \ \cancel{\text{furlong}}} \times \dfrac{3 \text{ ft}}{1 \ \cancel{\text{yd}}}$

$= 12 \times 220 \times 3 \text{ ft}$

$= 7,920 \text{ ft}$

43. 55 miles/hour

$= 55 \dfrac{\cancel{\text{mi}}}{\cancel{\text{hr}}} \times \dfrac{5,280 \text{ft}}{1 \ \cancel{\text{mi}}} \times \dfrac{1 \ \cancel{\text{hr}}}{60 \ \cancel{\text{min}}} \times \dfrac{1 \ \cancel{\text{min}}}{60 \text{ sec}}$

$= \dfrac{55 \times 5,280 \text{ ft}}{60 \times 60 \text{ sec}}$

$\approx 80.7 \text{ feet/second}$

45. 9.52 yards/second

$= 9.52 \dfrac{\cancel{\text{yds}}}{\cancel{\text{sec}}} \times \dfrac{3 \ \cancel{\text{ft}}}{1 \ \cancel{\text{yd}}} \times \dfrac{1 \text{ mi}}{5,280 \ \cancel{\text{ft}}} \times \dfrac{60 \ \cancel{\text{sec}}}{1 \ \cancel{\text{min}}}$

$\times \dfrac{60 \ \cancel{\text{min}}}{1 \text{ hr}}$

$= \dfrac{9.52 \times 3 \times 60 \times 60 \text{ mi}}{5,280 \text{ hr}}$

$\approx 19.5 \text{ miles per hour}$

47. 1,500 feet/second

$$= 1,500 \frac{\cancel{ft}}{\cancel{sec}} \times \frac{1mi}{5,280 \, \cancel{ft}} \times \frac{60 \, \cancel{sec}}{1 \, \cancel{min}} \times \frac{60 \, \cancel{min}}{1 \, hr}$$

$$= \frac{1,500 \times 60 \times 60 \, mi}{5,280 \, hr}$$

$$= 1,022.\overline{72} \, mi/hr$$

$$\approx 1,023 \, miles/hour$$

49.

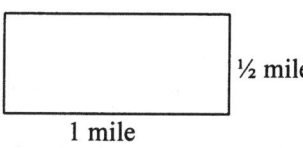

½ mile

1 mile

1st, find the perimeter:

Perimeter $= 2l + 2w$

$$= 2(1 \, mi) + 2\left(\frac{1}{2} mi\right)$$

$$= 2 \, mi + 1 \, mi$$

$$= 3 \, mi$$

Convert miles to dollars:

$$3 \, miles = 3 \, \cancel{mi} \times \frac{5,280 \, \cancel{ft}}{1 \, \cancel{mi}} \times \frac{\$1.15}{1 \, \cancel{ft}}$$

$$= 3 \times 5,280 \times 1.15 \, dollars$$

$$= \$18,216$$

It will cost $18,216.

51. 120 days $= 120 \, \cancel{days} \times \dfrac{1 \, \cancel{bale}}{8 \, \cancel{day}} \times \dfrac{\$10.50}{1 \, \cancel{bale}}$

$$= \frac{120 \times 10.50 \, dollars}{8}$$

$$= \$157.50$$

Calculator Problems

53. 751 miles $= 751 \, \cancel{mi} \times \dfrac{5,280 \, ft}{1 \, \cancel{mi}}$

$$= 751 \times 5,280 \, ft$$

$$= 3,965,280 \, ft$$

55. 4,982 yd $= 4,982 \, \cancel{yd} \times \dfrac{3 \, \cancel{ft}}{1 \, \cancel{yd}} \times \dfrac{12 \, in}{1 \, \cancel{ft}}$

$$= 4,982 \times 3 \times 12 \, in$$

$$= 179,352 \, in$$

57. 14,494 feet $= 14,494 \, \cancel{ft} \times \dfrac{1 \, mi}{5,280 \, \cancel{ft}}$

$$= \frac{14,494}{5,280} \, ft$$

$$\approx 2.7 \, mi$$

59. 3,427 miles $= 3,427 \, \cancel{mi} \times \dfrac{5,280 \, ft}{1 \, \cancel{mi}}$

$$= 3,427 \times 5,280 \, ft$$

$$= 18,094,560 \, ft$$

Review Problems

61. $\dfrac{2}{3} \cdot \dfrac{1}{2} = \dfrac{\cancel{2} \cdot 1}{3 \cdot \cancel{2}} = \dfrac{1}{3}$

63. $8\left(\dfrac{3}{4}\right) = \dfrac{2 \cdot \cancel{4} \cdot 3}{\cancel{4}} = 6$

65. $1\dfrac{1}{2} \cdot 2\dfrac{1}{3} = \dfrac{\cancel{3}}{2} \cdot \dfrac{7}{\cancel{3}} = \dfrac{7}{2} = 3\dfrac{1}{2}$

67. $\dfrac{3}{4} \div \dfrac{1}{8} = \dfrac{3}{4} \cdot \dfrac{8}{1} = \dfrac{3 \cdot 2 \cdot \cancel{4}}{\cancel{4} \cdot 1} = 6$

69. $4 \div \dfrac{2}{3} = 4 \cdot \dfrac{3}{2} = \dfrac{\cancel{2} \cdot 2 \cdot 3}{\cancel{2}} = 6$

71. $1\dfrac{3}{4} \div 2\dfrac{1}{2} = \dfrac{7}{4} \div \dfrac{5}{2} = \dfrac{7}{4} \cdot \dfrac{2}{5} = \dfrac{7 \cdot \cancel{2}}{\cancel{2} \cdot 2 \cdot 5} = \dfrac{7}{10}$

Improving Your Quantitative Literacy

73.   2,000 steps =

$$= 2{,}000 \ \cancel{\text{steps}} \ \times \frac{2.5 \ \cancel{\text{ft}}}{1 \ \cancel{\text{step}}} \times \frac{1 \ \text{mi}}{5{,}280 \ \cancel{\text{ft}}}$$

$$= \frac{2{,}000 \times 2.5}{5{,}280} \ \text{mi}$$

$$\approx 0.95 \ \text{mi} \approx 1 \ \text{mi}$$

10,000 steps =

$$= 10{,}000 \ \cancel{\text{steps}} \ \times \frac{2.5 \ \cancel{\text{ft}}}{1 \ \cancel{\text{step}}} \times \frac{1 \ \text{mi}}{5{,}280 \ \cancel{\text{ft}}}$$

$$= \frac{10{,}000 \times 2.5}{5{,}280} \ \text{mi}$$

$$\approx 4.73 \ \text{mi} \approx 5 \ \text{mi}$$

Both statements are correct.

Section 6.2

1. $3 \text{ ft}^2 = 3 \text{ ft}^2 \times \dfrac{144 \text{ in}^2}{1 \text{ ft}^2}$

$= 3 \times 144 \text{ in}^2$

$= 432 \text{ in}^2$

3. $288 \text{ in}^2 = 288 \text{ in}^2 \times \dfrac{1 \text{ ft}^2}{144 \text{ in}^2}$

$= \dfrac{288}{144} \text{ ft}^2$

$= 2 \text{ ft}^2$

5. $30 \text{ acres} = 30 \text{ acres} \times \dfrac{43,560 \text{ ft}^2}{1 \text{ acre}}$

$= 30 \times 43,560 \text{ ft}^2$

$= 1,306,800 \text{ ft}^2$

7. $2 \text{ mi}^2 = 2 \text{ mi}^2 \times \dfrac{640 \text{ acres}}{1 \text{ mi}^2}$

$= 2 \times 640 \text{ acres}$

$= 1,280 \text{ acres}$

9. $1,920 \text{ acres} = 1,920 \text{ acres} \times \dfrac{1 \text{ mi}^2}{640 \text{ acres}}$

$= \dfrac{1,920}{640} \text{mi}^2$

$= 3 \text{ mi}^2$

11. $12 \text{ yd}^2 = 12 \text{ yd}^2 \times \dfrac{9 \text{ ft}^2}{1 \text{ yd}^2}$

$= 12 \times 9 \text{ ft}^2$

$= 108 \text{ ft}^2$

13. $17 \text{ cm}^2 = 17 \text{ cm}^2 \times \dfrac{100 \text{ mm}^2}{1 \text{ cm}^2}$

$= 17 \times 100 \text{ mm}^2$

$= 1,700 \text{ mm}^2$

15. $2.8 \text{ m}^2 = 2.8 \text{ m}^2 \times \dfrac{100 \text{ dm}^2}{1 \text{ m}^2} \times \dfrac{100 \text{ cm}^2}{1 \text{ dm}^2}$

$= 2.8 \times 100 \times 100 \text{ cm}^2$

$= 28,000 \text{ cm}^2$

17. $1,200 \text{ mm}^2$

$= 1,200 \text{ mm}^2 \times \dfrac{1 \text{ cm}^2}{100 \text{ mm}^2} \times \dfrac{1 \text{ dm}^2}{100 \text{ cm}^2}$

$\times \dfrac{1 \text{ m}^2}{100 \text{ dm}^2}$

$= \dfrac{1,200}{100 \times 100 \times 100} \text{m}^2$

$= 0.0012 \text{ m}^2$

19. $5 \text{ a} = 5 \text{ a} \times \dfrac{100 \text{ m}^2}{1 \text{ a}}$

$= 5 \times 100 \text{ m}^2$

$= 500 \text{ m}^2$

21. $7 \text{ ha} = 7 \text{ ha} \times \dfrac{100 \text{ a}}{1 \text{ ha}}$

$= 7 \times 100 \text{ a}$

$= 700 \text{ a}$

23. $342 \text{ a} = 342 \text{ a} \times \dfrac{1 \text{ ha}}{100 \text{ a}}$

$= \dfrac{342}{100} \text{ ha}$

$= 3.42 \text{ ha}$

Applying the Concepts

25. NFL field:

$$A = lw$$

$$= (100)\left(53\frac{1}{3}\right)$$

$$= 100\left(\frac{160}{3}\right)$$

$$= \frac{16,000}{3}$$

$$= 5,333\frac{1}{3} \text{ yd}^2 \text{ ;}$$

$$5,333\frac{1}{3}\text{yd}^2$$

$$= 5,333\frac{1}{3}\cancel{\text{yd}^2} \times \frac{9 \cancel{\text{ft}^2}}{1 \cancel{\text{yd}^2}} \times \frac{1 \text{ acre}}{43,560 \cancel{\text{ft}^2}}$$

$$= \frac{5,333\frac{1}{3} \times 9}{43,560} \text{ acres}$$

$$\approx 1.10 \text{ acres}$$

The NFL field is approximately 1.10 acres.

Canadian field:

$$A = lw = (110)(65) = 7,150 \text{ yd}^2$$

$$7,150 \text{ yd}^2 = 7,150 \ \cancel{\text{yd}^2} \times \frac{9 \cancel{\text{ft}^2}}{1 \cancel{\text{yd}^2}} \times \frac{1 \text{ acre}}{43,560 \cancel{\text{ft}^2}}$$

$$= \frac{7,150 \times 9}{43,560} \text{ acres}$$

$$\approx 1.48 \text{ acres}$$

The Canadian field is approximately 1.48 acres.

25. (continued)
Arena Football:

$$A = lw$$

$$= (50)\left(28\frac{1}{3}\right)$$

$$= 50\left(\frac{85}{3}\right)$$

$$= 1,416\frac{2}{3} \text{ yd}^2$$

$$1,416\frac{2}{3}\text{yd}^2$$

$$= 1,416\frac{2}{3} \ \cancel{\text{yd}^2} \times \frac{9 \cancel{\text{ft}^2}}{1 \ \cancel{\text{yd}^2}} \times \frac{1 \text{ acre}}{43,560 \cancel{\text{ft}^2}}$$

$$= \frac{1,416\frac{2}{3} \times 9}{43,560} \text{ acres}$$

$$\approx 0.29 \text{ acres}$$

The Arena is approximately 0.29 acres.

27. $A = lw = (100 \text{ m})(30 \text{ m}) = 3,000 \text{ m}^2$

$$3,000 \text{ m}^2 = 3,000 \ \cancel{\text{m}^2} \times \frac{1 \text{ a}}{100 \ \cancel{\text{m}^2}}$$

$$= \frac{3,000}{100} \text{a}$$

$$= 30 \text{ a}$$

The area of the pool is 30 a.

29. Find the area of 1 brick:

$$A = (10 \text{ cm})(20 \text{ cm}) = 200 \text{ cm}^2$$

Convert the area of the patio to bricks:
110 m²

$$= 110 \cancel{\text{m}^2} \times \frac{100 \cancel{\text{dm}^2}}{1 \cancel{\text{m}^2}} \times \frac{100 \cancel{\text{cm}^2}}{1 \cancel{\text{dm}^2}} \times \frac{1 \text{ brick}}{200 \cancel{\text{cm}^2}}$$

$$= \frac{110 \times 100 \times 100}{200} \text{ bricks}$$

$$= 5,500 \text{ bricks}$$

It will take 5,500 bricks to make the patio.

31. $5 \text{ yd}^3 = 5 \text{ yd}^3 \times \dfrac{27 \text{ ft}^3}{1 \text{ yd}^3}$

$= 5 \times 27 \text{ ft}^3$

$= 135 \text{ ft}^3$

33. $3 \text{ pt} = 3 \text{ pt} \times \dfrac{16 \text{ fl oz}}{1 \text{ pt}}$

$= 3 \times 16 \text{ fl oz}$

$= 48 \text{ fl oz}$

35. $2 \text{ gal} = 2 \text{ gal} \times \dfrac{4 \text{ qt}}{1 \text{ gal}}$

$= 2 \times 4 \text{ qt}$

$= 8 \text{ qt}$

37. $2.5 \text{ gal} = 2.5 \text{ gal} \times \dfrac{4 \text{ qt}}{1 \text{ gal}} \times \dfrac{2 \text{ pt}}{1 \text{ qt}}$

$= 2.5 \times 4 \times 2 \text{ pt}$

$= 20 \text{ pt}$

39. $15 \text{ qt} = 15 \text{ qt} \times \dfrac{2 \text{ pt}}{1 \text{ qt}} \times \dfrac{16 \text{ fl oz}}{1 \text{ pt}}$

$= 15 \times 2 \times 16 \text{ fl oz}$

$= 480 \text{ fl oz}$

41. $64 \text{ pt} = 64 \text{ pt} \times \dfrac{1 \text{ qt}}{2 \text{ pt}} \times \dfrac{1 \text{ gal}}{4 \text{ qt}}$

$= \dfrac{64}{2 \times 4} \text{ gal}$

$= 8 \text{ gal}$

43. $12 \text{ pt} = 12 \text{ pt} \times \dfrac{1 \text{ qt}}{2 \text{ pt}}$

$= \dfrac{12}{2} \text{ qt}$

$= 6 \text{ qt}$

45. $243 \text{ ft}^3 = 243 \text{ ft}^3 \times \dfrac{1 \text{ yd}^3}{27 \text{ ft}^3}$

$= \dfrac{243}{27} \text{ yd}^3$

$= 9 \text{ yd}^3$

47. $5 \text{ L} = 5 \text{ L} \times \dfrac{1,000 \text{ mL}}{1 \text{ L}}$

$= 5 \times 1,000 \text{ mL}$

$= 5,000 \text{ mL}$

49. $127 \text{ mL} = 127 \text{ mL} \times \dfrac{1 \text{ L}}{1,000 \text{ mL}}$

$= \dfrac{127}{1,000} \text{ L}$

$= 0.127 \text{ L}$

51. $4 \text{ kL} = 4 \text{ kL} \times \dfrac{1,000 \text{ L}}{1 \text{ kL}} \times \dfrac{1,000 \text{ mL}}{1 \text{ L}}$

$= 4 \times 1,000 \times 1,000 \text{ mL}$

$= 4,000,000 \text{ mL}$

53. $14.92 \text{ kL} = 14.92 \text{ kL} \times \dfrac{1,000 \text{ L}}{1 \text{ kL}}$

$= 14.92 \times 1,000 \text{ L}$

$= 14,920 \text{ L}$

Applying the Concepts

55. Convert 1 gallon to pints, then cups:

$1 \text{ gal} = 1 \text{ gal} \times \dfrac{4 \text{ qt}}{1 \text{ gal}} \times \dfrac{2 \text{ pt}}{1 \text{ qt}} \times \dfrac{1 \text{ cup}}{\frac{1}{2} \text{ pt}}$

$= \dfrac{4 \times 2}{\frac{1}{2}} \text{ cups}$

$= 16 \text{ cups}$

The coffee make can fill 16 cups.

57. $20 \text{ ft}^3 = 20 \text{ ft}^3 \times \dfrac{1,728 \text{ in}^3}{1 \text{ ft}^3}$

$= 20 \times 1,728 \text{ in}^3$

$= 34,560 \text{ in}^3$

59. Convert gallons to ounces, then glasses:
3 gal

$= 3 \text{ gal} \times \dfrac{4 \text{ qt}}{1 \text{ gal}} \times \dfrac{2 \text{ pt}}{1 \text{ qt}} \times \dfrac{16 \text{ fl oz}}{1 \text{ pt}} \times \dfrac{1 \text{ glass}}{8 \text{ fl oz}}$

$= \dfrac{3 \times 4 \times 2 \times 16}{8} \text{ glasses}$

$= 48 \text{ glasses}$

Calculator Problems

61. $31,700 \text{ mi}^2 = 31,700 \text{ mi}^2 \times \dfrac{640 \text{ acres}}{1 \text{ mi}^2}$

$= 31,700 \times 640 \text{ acres}$

$= 20,288,000 \text{ acres}$

63. $2,067,795 \text{ acres}$

$= 2,067,795 \text{ acres} \times \dfrac{1 \text{ mi}^2}{640 \text{ acres}}$

$= \dfrac{2,067,795}{640} \text{ mi}^2$

$= 3,230.93 \text{ mi}^2$

65. $93.4 \text{ qt} = 93.4 \text{ qt} \times \dfrac{1 \text{ gal}}{4 \text{ qt}}$

$= \dfrac{93.4}{4} \text{ gal}$

$= 23.35 \text{ gal}$

67. $796 \text{ yd}^3 = 796 \text{ yd}^3 \times \dfrac{27 \text{ ft}^3}{1 \text{ yd}^3}$

$= 796 \times 27 \text{ ft}^3$

$= 21,492 \text{ ft}^3$

69. $10,585,000 \text{ yd}^3 = 10,585,000 \text{ yd}^3 \times \dfrac{27 \text{ ft}^3}{1 \text{ yd}^3}$

$= 10.585,000 \times 27 \text{ ft}^3$

$= 285,795,000 \text{ ft}^3$

Review Problems

71. $\dfrac{3}{8} + \dfrac{1}{4} = \dfrac{3}{8} + \dfrac{1 \cdot 2}{4 \cdot 2} = \dfrac{3}{8} + \dfrac{2}{8} = \dfrac{3+2}{8} = \dfrac{5}{8}$

73. $3\dfrac{1}{2}$

$+ 5\dfrac{1}{2}$

$8\dfrac{2}{2} = 8 + 1 = 9$

75. $\dfrac{7}{15} - \dfrac{2}{15} = \dfrac{7-2}{15} = \dfrac{5}{15} = \dfrac{1}{3}$

77. $\left.\begin{array}{l} 36 = 2 \cdot 2 \cdot 3 \cdot 3 \\ 48 = 2 \cdot 2 \cdot 2 \cdot 2 \cdot 3 \end{array}\right\} \text{LCD} = 2 \cdot 2 \cdot 2 \cdot 2 \cdot 3 \cdot 3 = 144$

$\dfrac{5}{36} - \dfrac{1}{48} = \dfrac{5 \cdot 4}{36 \cdot 4} - \dfrac{1 \cdot 3}{48 \cdot 4} = \dfrac{20}{144} - \dfrac{3}{144} = \dfrac{17}{144}$

79. $5\dfrac{1}{6} = \quad 5\dfrac{1 \cdot 3}{6 \cdot 3} = \quad 5\dfrac{3}{18}$

$- 2\dfrac{1}{9} = - 2\dfrac{1 \cdot 2}{9 \cdot 2} = - 2\dfrac{2}{18}$

$3\dfrac{1}{18}$

Section 6.3

1. $8 \text{ lb} = 8 \text{ lb} \times \dfrac{16 \text{ oz}}{1 \text{ lb}}$

$= 8 \times 16 \text{ oz}$

$= 128 \text{ oz}$

3. $2 \text{ T} = 2 \text{ T} \times \dfrac{2{,}000 \text{ lb}}{1 \text{ T}}$

$= 2 \times 2{,}000 \text{ lb}$

$= 4{,}000 \text{ lb}$

5. $192 \text{ oz} = 192 \text{ oz} \times \dfrac{1 \text{ lb}}{16 \text{ oz}}$

$= \dfrac{192}{16} \text{lb}$

$= 12 \text{ lb}$

7. $1{,}800 \text{ lb} = 1{,}800 \text{ lb} \times \dfrac{1 \text{ T}}{2{,}000 \text{ lb}}$

$= \dfrac{1{,}800}{2{,}000} \text{T}$

$= 0.9 \text{ T}$

9. $1 \text{ T} = 1 \text{ T} \times \dfrac{2{,}000 \text{ lb}}{1 \text{ T}} \times \dfrac{16 \text{ oz}}{1 \text{ lb}}$

$= 1 \times 2{,}000 \times 16 \text{ oz}$

$= 32{,}000 \text{ oz}$

11. $3\dfrac{1}{2} \text{ lb} = 3\dfrac{1}{2} \text{ lb} \times \dfrac{16 \text{ oz}}{1 \text{ lb}}$

$= 3\dfrac{1}{2} \times 16 \text{ oz}$

$= \dfrac{7}{2} \times 16 \text{ oz}$

$= 56 \text{ oz}$

13. $6\dfrac{1}{2} \text{ T} = 6\dfrac{1}{2} \text{ T} \times \dfrac{2{,}000 \text{ lb}}{1 \text{ T}}$

$= \dfrac{13}{2} \times 2{,}000 \text{ lb}$

$= 13{,}000 \text{ lb}$

15. $2 \text{ kg} = 2 \text{ kg} \times \dfrac{1{,}000 \text{ g}}{1 \text{ kg}}$

$= 2 \times 1{,}000 \text{ g}$

$= 2{,}000 \text{ g}$

17. $4 \text{ cg} = 4 \text{ cg} \times \dfrac{1 \text{ g}}{100 \text{ cg}} \times \dfrac{1{,}000 \text{ mg}}{1 \text{ g}}$

$= \dfrac{4 \times 1{,}000}{100} \text{ mg}$

$= 40 \text{ mg}$

19. $2 \text{ kg} = 2 \text{ kg} \times \dfrac{1{,}000 \text{ g}}{1 \text{ kg}} \times \dfrac{100 \text{ cg}}{1 \text{ g}}$

$= 2 \times 1{,}000 \times 100 \text{ cg}$

$= 200{,}000 \text{ cg}$

21. $5.08 \text{ g} = 5.08 \text{ g} \times \dfrac{100 \text{ cg}}{1 \text{ g}}$

$= 5.08 \times 100 \text{ cg}$

$= 508 \text{ cg}$

23. $450 \text{ cg} = 450 \text{ cg} \times \dfrac{1 \text{ g}}{100 \text{ cg}}$

$= \dfrac{450}{100} \text{ g}$

$= 4.5 \text{ g}$

25. 478.95 mg

$$= 478.95 \text{ mg} \times \frac{1 \text{ g}}{1,000 \text{ mg}} \times \frac{100 \text{ cg}}{1 \text{ g}}$$

$$= \frac{478.95 \times 100}{1,000} \text{ cg}$$

$$= 47.895 \text{ cg}$$

27. $1,578 \text{ mg} = 1,578 \text{ mg} \times \frac{1 \text{ g}}{1,000 \text{ mg}}$

$$= \frac{1,578}{1,000} \text{ g}$$

$$= 1.578 \text{ g}$$

29. $42,000 \text{ cg} = 42,000 \text{ cg} \times \frac{1 \text{ g}}{100 \text{ cg}} \times \frac{1 \text{ kg}}{1,000 \text{ g}}$

$$= \frac{42,000}{100 \times 1,000} \text{kg}$$

$$= 0.42 \text{ kg}$$

31. First, find the total mg in the bottle:

$$\text{Total mg} = 60 \text{ softgels} \cdot \frac{800 \text{ mg}}{\text{softgel}} = 48,000 \text{ mg}$$

Then convert mg to g:

$$48,000 \text{ mg} = 48,000 \text{ mg} \times \frac{1 \text{ g}}{1,000 \text{ mg}}$$

$$= \frac{48,000}{1,000} \text{ g}$$

$$= 48 \text{ g}$$

The bottle contains 48 grams of fatty acid.

33. First, find the total mg in the bottle:

$$\text{Total mg} = 80 \text{ vitamins} \cdot \frac{50 \text{ mg}}{\text{vitamin}} = 4,000 \text{ mg}$$

Then convert mg to g:

$$4,000 \text{ mg} = 4,000 \text{ mg} \times \frac{1 \text{ g}}{1,000 \text{ mg}}$$

$$= \frac{4,000}{1,000} \text{ g}$$

$$= 4 \text{ g}$$

The bottle contains 4 grams of Riboflavin.

35. First, find the total mg in the bottle:

$$\text{Total mg} = 120 \text{ tablets} \cdot \frac{81 \text{ mg}}{\text{tablet}} = 9,720 \text{ mg}$$

Then convert mg to g:

$$9,720 \text{ mg} = 9,720 \text{ mg} \times \frac{1 \text{ g}}{1,000 \text{ mg}}$$

$$= \frac{9,720}{1,000} \text{ g}$$

$$= 9.72 \text{ g}$$

The bottle contains 9.72 grams of aspirin.

37. First, find the total mg in the bottle:

$$\text{Total mg} = 240 \text{ vitamins} \cdot \frac{500 \text{ mg}}{\text{vitamin}} = 120,000 \text{ mg}$$

Then convert mg to g:

$$120,000 \text{ mg} = 120,000 \text{ mg} \times \frac{1 \text{ g}}{1,000 \text{ mg}}$$

$$= \frac{120,000}{1,000} \text{ g}$$

$$= 120 \text{ g}$$

The bottle contains 120 grams of Vitamin C.

39. First, find the total mL in a six pack:

$$\text{Total ml} = 6 \text{ bottles} \cdot \frac{500 \text{ mL}}{\text{bottle}} = 3,000 \text{ mL}$$

Then convert mL to L

$$3,000 \text{ mL} = 3,000 \text{ mL} \times \frac{1 \text{ L}}{1,000 \text{ mL}}$$

$$= \frac{3,000}{1,000} \text{ L}$$

$$= 3 \text{ L}$$

There are 3 L in a six pack.

51.

$$\begin{array}{r} 0.125 \\ 8\overline{)1.000} \\ \underline{8\downarrow} \\ 20 \\ \underline{16\downarrow} \\ 40 \\ \underline{40} \\ 0 \end{array} \qquad \frac{1}{8} = 0.125$$

41. First, find the total mL in a six pack:

$$\text{Total ml} = 6 \text{ bottles} \cdot \frac{250 \text{ mL}}{\text{bottle}} = 1,500 \text{ mL}$$

Then convert mL to L

$$1,500 \text{ mL} = 1,500 \text{ mL} \times \frac{1 \text{ L}}{1,000 \text{ mL}}$$

$$= \frac{1,500}{1,000} \text{ L}$$

$$= 1.5 \text{ L}$$

There are 1.5 L in a six pack.

Review Problems

43. $\quad 0.18 = \dfrac{18}{100} = \dfrac{9}{50}$

45. $\quad 0.09 = \dfrac{9}{100}$

47. $\quad 0.8 = \dfrac{8}{10} = \dfrac{4}{5}$

49.

$$\begin{array}{r} 0.9 \\ 10\overline{)9.0} \\ \underline{90} \\ 0 \end{array} \qquad \frac{9}{10} = 0.9$$

Section 6.4

1. $6 \text{ in} = 6 \text{ in} \times \dfrac{2.54 \text{ cm}}{1 \text{ in}}$

$= 6 \times 2.54 \text{ cm}$

$= 15.24 \text{ cm}$

3. $4 \text{ m} = 4 \text{ m} \times \dfrac{3.28 \text{ ft}}{1 \text{ m}}$

$= 4 \times 3.28 \text{ ft}$

$= 13.12 \text{ ft}$

5. $6 \text{ m} = 6 \text{ m} \times \dfrac{3.28 \text{ ft}}{1 \text{ m}} \times \dfrac{1 \text{ yd}}{3 \text{ ft}}$

$= \dfrac{6 \times 3.28}{3} \text{ yd}$

$= 6.56 \text{ yd}$

7. $20 \text{ mi} = 20 \text{ mi} \times \dfrac{1.61 \text{ km}}{1 \text{ mi}} \times \dfrac{1,000 \text{ m}}{1 \text{ km}}$

$= 20 \times 1.61 \times 1,000 \text{ m}$

$= 32,200 \text{ m}$

9. $5 \text{ m}^2 = 5 \text{ m}^2 \times \dfrac{1.196 \text{ yd}^2}{1 \text{ m}^2}$

$= 5 \times 1.196 \text{ yd}^2$

$= 5.98 \text{ yd}^2$

11. $10 \text{ ha} = 10 \text{ ha} \times \dfrac{2.47 \text{ acres}}{1 \text{ ha}}$

$= 10 \times 2.47 \text{ acres}$

$= 24.7 \text{ acres}$

13. $500 \text{ in}^3 = 500 \text{ in}^3 \times \dfrac{16.39 \text{ mL}}{1 \text{ in}^3}$

$= 500 \times 16.39 \text{ mL}$

$= 8,195 \text{ mL}$

15. $2 \text{ L} = 2 \text{ L} \times \dfrac{1.06 \text{ qt}}{1 \text{ L}}$

$= 2 \times 1.06 \text{ qt}$

$= 2.12 \text{ qt}$

17. $20 \text{ gal} = 20 \text{ gal} \times \dfrac{3.79 \text{ L}}{1 \text{ gal}}$

$= 20 \times 3.79 \text{ L}$

$= 75.8 \text{ L}$

19. $12 \text{ oz} = 12 \text{ oz} \times \dfrac{28.3 \text{ g}}{1 \text{ oz}}$

$= 12 \times 28.3 \text{ g}$

$= 339.6 \text{ g}$

21. $15 \text{ kg} = 15 \text{ kg} \times \dfrac{2.20 \text{ lb}}{1 \text{ kg}}$

$= 15 \times 2.20 \text{ lb}$

$= 33 \text{ lb}$

23. When $C = 185$:

$F = \dfrac{9}{5} C + 32$

$= \dfrac{9}{5}(185) + 32$

$= 333 + 32$

$= 365$

185°C is equivalent to 365°F.

25. When $F = 86$:

$C = \dfrac{5(F - 32)}{9}$

$= \dfrac{5(86 - 32)}{9}$

$= \dfrac{5(54)}{9}$

$= 30$

86°F is equivalent to 30°C.

Calculator Problems

27. $10 \text{ cm} = 10 \text{ cm} \times \dfrac{1 \text{ in}}{2.54 \text{ cm}}$

$\qquad = \dfrac{10}{2.54} \text{ in}$

$\qquad \approx 3.94 \text{ in}$

29. $25 \text{ ft} = 25 \text{ ft} \times \dfrac{1 \text{ m}}{3.28 \text{ ft}}$

$\qquad = \dfrac{25}{3.28} \text{ m}$

$\qquad \approx 7.62 \text{ m}$

31. $49 \text{ qt} = 49 \text{ qt} \times \dfrac{1 \text{ L}}{1.06 \text{ qt}}$

$\qquad = \dfrac{49}{1.06} \text{ L}$

$\qquad \approx 46.23 \text{ L}$

33. $500 \text{ g} = 500 \text{ g} \times \dfrac{1 \text{ oz}}{28.3 \text{ g}}$

$\qquad = \dfrac{500}{28.3} \text{ oz}$

$\qquad \approx 17.67 \text{ oz}$

35. Answers will vary.

37. $100 \text{ yd} = 100 \text{ yd} \times \dfrac{3 \text{ ft}}{1 \text{ yd}} \times \dfrac{1 \text{ m}}{3.28 \text{ ft}}$

$\qquad = \dfrac{100 \times 3}{3.28} \text{ m}$

$\qquad \approx 91.46 \text{ m}$

39. $25 \text{ yd}^2 = 25 \text{ yd}^2 \times \dfrac{1 \text{ m}^2}{1.196 \text{ yd}^2}$

$\qquad = \dfrac{25}{1.196} \text{ m}^2$

$\qquad \approx 20.90 \text{ m}^2$

41. $55 \text{ miles per hour} = 55 \dfrac{\text{mi}}{\text{hr}} \times \dfrac{1.61 \text{ km}}{1 \text{ mi}}$

$\qquad = 55 \times 1.61 \dfrac{\text{km}}{\text{hr}}$

$\qquad = 88.55 \dfrac{\text{km}}{\text{hr}}$

43. First convert 8 in to ft:

$8 \text{ in} = 8 \text{ in} \times \dfrac{1 \text{ ft}}{12 \text{ in}}$

$\qquad = \dfrac{8}{12} \text{ ft}$

$\qquad = \dfrac{2}{3} \text{ ft}$

$6 \text{ ft } 8 \text{ in} = 6\dfrac{2}{3} \text{ ft}$

Second, convert ft to m:

$6\dfrac{2}{3} \text{ ft} = 6\dfrac{2}{3} \text{ ft} \times \dfrac{1 \text{ m}}{3.28 \text{ ft}}$

$\qquad = \dfrac{6\frac{2}{3}}{3.28} \text{ m}$

$\qquad \approx 2.03 \text{ m}$

6 ft 8 in is equivalent to 2.03 m.

45. When $F = 101°$:

$C = \dfrac{5(F - 32)}{9}$

$\quad = \dfrac{5(101 - 32)}{9}$

$\quad = \dfrac{5(69)}{9}$

$\quad \approx 38.3$

101°F is equivalent to 38.3°C.

Review Problems

47.

$$4\overline{)3.00}^{\,0.75}$$
$$\underline{28}\downarrow$$
$$20$$
$$\underline{20}$$
$$0$$

$$\frac{3}{4} = 0.75$$

49. $5\dfrac{1}{2} = \dfrac{11}{2}:$

$$2\overline{)11.0}^{\,5.5}$$
$$\underline{10}\downarrow$$
$$10$$
$$\underline{10}$$
$$0$$

$$5\frac{1}{2} = 5.5$$

51. $\dfrac{3}{100} = 0.03$

53. $0.34 = \dfrac{34}{100} = \dfrac{17}{50}$

55. $2.4 = 2\dfrac{4}{10} = 2\dfrac{2}{5}$

57. $1.75 = 1\dfrac{75}{100} = 1\dfrac{3}{4}$

Section 6.5

1. a. $4 \text{ hr } 30 \text{ min} = 4 \text{ hr} \times \dfrac{60 \text{ min}}{1 \text{ hr}} + 30 \text{ min}$
$= 240 \text{ min} + 30 \text{ min}$
$= 270 \text{ min}$

b. $4 \text{ hr } 30 \text{ min} = 4 \text{ hr} + 30 \text{ min} \times \dfrac{1 \text{ hr}}{60 \text{ min}}$
$= 4 \text{ hr} + 0.5 \text{ hr}$
$= 4.5 \text{ hr}$

3. a. $5 \text{ hr } 20 \text{ min} = 5 \text{ hr} \times \dfrac{60 \text{ min}}{1 \text{ hr}} + 20 \text{ min}$
$= 300 \text{ min} + 20 \text{ min}$
$= 320 \text{ min}$

b. $5 \text{ hr } 20 \text{ min} = 5 \text{ hr} + 20 \text{ min} \times \dfrac{1 \text{ hr}}{60 \text{ min}}$
$= 5 \text{ hr} + 0.333 \text{ hr}$
$= 5.333 \text{ hr}$

5. a. $6 \text{ min } 30 \text{ sec} = 6 \text{ min} \times \dfrac{60 \text{ sec}}{1 \text{ min}} + 30 \text{ sec}$
$= 360 \text{ sec} + 30 \text{ sec}$
$= 390 \text{ sec}$

b. $6 \text{ min } 30 \text{ sec} = 6 \text{ min} + 30 \text{ sec} \times \dfrac{1 \text{ min}}{60 \text{ sec}}$
$= 6 \text{ min} + 0.5 \text{ min}$
$= 6.5 \text{ min}$

7. a. $5 \text{ min } 20 \text{ sec} = 5 \text{ min} \times \dfrac{60 \text{ sec}}{1 \text{ min}} + 20 \text{ sec}$
$= 300 \text{ sec} + 20 \text{ sec}$
$= 320 \text{ sec}$

b. $5 \text{ min } 20 \text{ sec} = 5 \text{ min} + 20 \text{ sec} \times \dfrac{1 \text{ min}}{60 \text{ sec}}$
$= 5 \text{ min} + 0.33 \text{ min}$
$= 5.33 \text{ min}$

9. a. $2 \text{ lbs } 8 \text{ oz} = 2 \text{ lb} \times \dfrac{16 \text{ oz}}{1 \text{ lb}} + 8 \text{ oz}$
$= 32 \text{ oz} + 8 \text{ oz}$
$= 40 \text{ oz}$

b. $2 \text{ lb } 8 \text{ oz} = 2 \text{ lb} + 8 \text{ oz} \times \dfrac{1 \text{ lb}}{16 \text{ oz}}$
$= 2 \text{ lb} + 0.5 \text{ lb}$
$= 2.5 \text{ lb}$

11. a. $4 \text{ lbs } 12 \text{ oz} = 4 \text{ lb} \times \dfrac{16 \text{ oz}}{1 \text{ lb}} + 12 \text{ oz}$
$= 64 \text{ oz} + 12 \text{ oz}$
$= 76 \text{ oz}$

b. $4 \text{ lb } 12 \text{ oz} = 4 \text{ lb} + 12 \text{ oz} \times \dfrac{1 \text{ lb}}{16 \text{ oz}}$
$= 4 \text{ lb} + 0.75 \text{ lb}$
$= 4.75 \text{ lb}$

13. a. $4 \text{ ft } 6 \text{ in} = 4 \text{ ft} \times \dfrac{12 \text{ in}}{1 \text{ ft}} + 6 \text{ in}$
$= 48 \text{ in} + 6 \text{ in}$
$= 54 \text{ in}$

b. $4 \text{ ft } 6 \text{ in} = 4 \text{ ft} + 6 \text{ in} \times \dfrac{1 \text{ ft}}{12 \text{ in}}$
$= 4 \text{ ft} + 0.5 \text{ ft}$
$= 4.5 \text{ ft}$

15. a. $5 \text{ ft } 9 \text{ in} = 5 \text{ ft} \times \dfrac{12 \text{ in}}{1 \text{ ft}} + 9 \text{ in}$
$= 60 \text{ in} + 9 \text{ in}$
$= 69 \text{ in}$

b. $5 \text{ ft } 9 \text{ in} = 5 \text{ ft} + 9 \text{ in} \times \dfrac{1 \text{ ft}}{12 \text{ in}}$
$= 5 \text{ ft} + 0.75 \text{ ft}$
$= 5.75 \text{ ft}$

17. a. $2 \text{ gal } 1 \text{ qt} = 2 \text{ } \cancel{\text{gal}} \times \dfrac{4 \text{ qt}}{1 \text{ } \cancel{\text{gal}}} + 1 \text{ qt}$

$\qquad\qquad = 8 \text{ qt} + 1 \text{ qt}$

$\qquad\qquad = 9 \text{ qt}$

b. $2 \text{ gal } 1 \text{ qt} = 2 \text{ gal} + 1 \text{ } \cancel{\text{qt}} \times \dfrac{1 \text{ gal}}{4 \text{ } \cancel{\text{qt}}}$

$\qquad\qquad = 2 \text{ gal} + 0.25 \text{ gal}$

$\qquad\qquad = 2.25 \text{ gal}$

19.
$$
\begin{array}{r}
4 \text{ hr } 47 \text{ min} \\
+ \ 6 \text{ hr } 13 \text{ min} \\
\hline
\end{array}
$$
10 hr 60 min = 10 hr + 1 hr = 11 hr

21.
$$
\begin{array}{r}
8 \text{ ft } 10 \text{ in} \\
+ \ 13 \text{ ft } \ 6 \text{ in} \\
\hline
\end{array}
$$
21 ft 16 in = 21 ft + 1 ft 4 in
$\qquad\qquad\quad$ = 22 ft 4 in

23.
$$
\begin{array}{r}
4 \text{ lb } 12 \text{ oz} \\
+ \ 6 \text{ lb } \ 4 \text{ oz} \\
\hline
\end{array}
$$
10 lb 16 oz = 10 lb + 1 lb = 11 lb

25.
$$
\begin{array}{r}
8 \text{ hr } 15 \text{ min} = \quad 7 \text{ hr } 75 \text{ min} \\
- \ 2 \text{ hr } 35 \text{ min} = - \ 2 \text{ hr } 35 \text{ min} \\
\hline
5 \text{ hr } 40 \text{ min}
\end{array}
$$

27.
$$
\begin{array}{r}
7 \text{ hr } 30 \text{ min} = \quad 6 \text{ hr } 90 \text{ min} \\
- \ 3 \text{ hr } 43 \text{ min} = - \ 3 \text{ hr } 43 \text{ min} \\
\hline
3 \text{ hr } 47 \text{ min}
\end{array}
$$

29.
$$
\begin{array}{r}
5 \text{ hr } \ 9 \text{ min} = \quad 4 \text{ hr } 69 \text{ min} \\
- \ 4 \text{ hr } 17 \text{ min} = - \ 4 \text{ hr } 17 \text{ min} \\
\hline
52 \text{ min}
\end{array}
$$

Applying the Concepts

31.
$$
\begin{array}{r}
50 \text{ min } 36 \text{ sec} \\
4 \text{ hr } 40 \text{ min } \ 4 \text{ sec} \\
+ \ 2 \text{ hr } 47 \text{ min } 38 \text{ sec} \\
\hline
6 \text{ hr } 137 \text{ min } 78 \text{ sec}
\end{array}
$$
= 6 hr + 2 hr 17 min + 1 min 18 sec
= 8 hr 18 min 18 sec
Reid's total time is 8:18:18

$$
\begin{array}{r}
56 \text{ min } \ 51 \text{ sec} \\
5 \text{ hr } \ 9 \text{ min} \\
+ \ 3 \text{ hr } \ 2 \text{ min } 10 \text{ sec} \\
\hline
8 \text{ hr } 67 \text{ min } 61 \text{ sec}
\end{array}
$$
= 8 hr + 1 hr 7 min + 1 min 1 sec
= 9 hr 8 min 1sec
Bowden's total time is 9:08:01

33.
$$
\begin{array}{r}
56 \text{ min } 51 \text{ sec} \\
- \ 50 \text{ min } \ 36 \text{ sec} \\
\hline
6 \text{ min } \ 15 \text{ sec}
\end{array}
$$
Peter was 00:06:15 faster.

35. Total weight:
$$
\begin{array}{r}
6 \text{ lb } 8 \text{ oz} \\
\times \qquad\quad 4 \\
\hline
24 \text{ lb } 32 \text{ oz}
\end{array}
$$

$24 \text{ lb } 32 \text{ oz} = 24 \text{ lb} + 32 \text{ oz} \times \dfrac{1 \text{ lb}}{16 \text{ oz}}$

$\qquad\qquad\qquad = 24 \text{ lb} + 2 \text{ lb}$

$\qquad\qquad\qquad = 26 \text{ lb}$

Total cost: $26 \text{ } \cancel{\text{lb}} \times \dfrac{\$4.00}{\cancel{\text{lb}}} = \$104$

37. Time per week:
$$
\begin{array}{r}
1 \text{ hr } 15 \text{ min} \\
\times \qquad\quad 4 \\
\hline
4 \text{ hr } 60 \text{ min} = 5 \text{ hr}
\end{array}
$$

Total time: $5 \dfrac{\text{hr}}{\cancel{\text{wk}}} \times 2 \text{ } \cancel{\text{wk}} = 10 \text{ hr}$

39. Total fabric: 3 yd 1 ft

$$\begin{array}{r} 3 \text{ yd }1\text{ ft} \\ \times \qquad 6 \\ \hline 18 \text{ yd }6\text{ ft} \end{array}$$

$$18 \text{ yd }6\text{ ft} = 18 \text{ yd} + 6 \ \cancel{\text{ft}} \times \frac{1 \text{ yd}}{3 \ \cancel{\text{ft}}}$$

$$= 18 \text{ yd} + 2 \text{ yd}$$

$$= 20 \text{ yd}$$

Total cost: $20 \ \cancel{\text{yd}} \ \times \dfrac{\$7.50}{\cancel{\text{yd}}} = \$150$

41. Total weight of 6 avocados (in lbs)

$$6 \ \cancel{\text{avoc}} \times \frac{8 \ \cancel{\text{oz}}}{\cancel{\text{avoc}}} \times \frac{1 \text{ lb}}{16 \ \cancel{\text{oz}}} = \frac{6 \times 8}{16} \text{ lb} = 3 \text{ lb}$$

Total cost: $3 \ \cancel{\text{lb}} \ \times \dfrac{\$2.00}{\cancel{\text{lb}}} = \$6$

Review Problems

43. $\left(\dfrac{1}{2}\right)^3 = \dfrac{1}{2} \cdot \dfrac{1}{2} \cdot \dfrac{1}{2} = \dfrac{1}{8}$

45. $\left(2\dfrac{1}{2}\right)^2 = \left(\dfrac{5}{2}\right)^2 = \dfrac{5}{2} \cdot \dfrac{5}{2} = \dfrac{25}{4} = 6\dfrac{1}{4}$

47. $(0.5)^3 = (0.5)(0.5)(0.5) = 0.125$

49. $(2.5)^2 = (2.5)(2.5) = 6.25$

Extending the Concepts

51.

$$\begin{array}{ll} 2 \text{ min } 1.19 \text{ sec} = & 1 \text{ min } 61.19 \text{ sec} \\ - \ 1 \text{ min } 59.4 \text{ sec} = & - \ 1 \text{ min } 59.40 \text{ sec} \\ \hline & 1.79 \text{ sec} \end{array}$$

Secretariat was 1.79 seconds faster.

Chapter 6 Review

1. $12 \text{ ft} = 12 \cancel{\text{ft}} \times \dfrac{12 \text{ in}}{1 \cancel{\text{ft}}}$

$= 12 \times 12 \text{ in}$

$= 144 \text{ in}$

3. $49 \text{ cm} = 49 \cancel{\text{cm}} \times \dfrac{1 \text{ m}}{100 \cancel{\text{cm}}}$

$= \dfrac{49}{100} \text{m}$

$= 0.49 \text{ m}$

5. $10 \text{ acres} = 10 \cancel{\text{acres}} \times \dfrac{43,560 \text{ ft}^2}{1 \cancel{\text{acre}}}$

$= 10 \times 43,560 \text{ ft}^2$

$= 435,600 \text{ ft}^2$

7. $4 \text{ ft}^2 = 4 \cancel{\text{ft}^2} \times \dfrac{144 \text{ in}^2}{1 \cancel{\text{ft}^2}}$

$= 4 \times 144 \text{ in}^2$

$= 576 \text{ in}^2$

9. $24 \text{ qt} = 24 \cancel{\text{qt}} \times \dfrac{1 \text{ gal}}{4 \cancel{\text{qt}}}$

$= \dfrac{24}{4} \text{gal}$

$= 6 \text{ gal}$

11. $8 \text{ lb} = 8 \cancel{\text{lb}} \times \dfrac{16 \text{ oz}}{1 \cancel{\text{lb}}}$

$= 8 \times 16 \text{ oz}$

$= 128 \text{ oz}$

13. $5 \text{ kg} = 5 \cancel{\text{kg}} \times \dfrac{1,000 \text{ g}}{1 \cancel{\text{kg}}}$

$= 5 \times 1,000 \text{ g}$

$= 5,000 \text{ g}$

15. $4 \text{ in} = 4 \cancel{\text{in}} \times \dfrac{2.54 \text{ cm}}{1 \cancel{\text{in}}}$

$= 4 \times 2.54 \text{ cm}$

$= 10.16 \text{ cm}$

17. $7 \text{ L} = 7 \cancel{\text{L}} \times \dfrac{1.06 \text{ qts}}{1 \cancel{\text{L}}}$

$= 7 \times 1.06 \text{ qt}$

$= 7.42 \text{ qt}$

19. $5 \text{ oz} = 5 \cancel{\text{oz}} \times \dfrac{28.3 \text{ g}}{1 \cancel{\text{oz}}}$

$= 5 \times 28.3 \text{ g}$

$= 141.5 \text{ g}$

21. When $C = 120$:

$F = \dfrac{9}{5}C + 32$

$= \dfrac{9}{5}(120) + 32$

$= 216 + 32$

$= 248$

120°C is equivalent to 248°F.

23. $24 \text{ cans} = 24 \text{ cans} \times \dfrac{355 \cancel{\text{mL}}}{1 \cancel{\text{can}}} \times \dfrac{1 \text{ L}}{1,000 \cancel{\text{mL}}}$

$= \dfrac{24 \times 355}{1,000} \text{ L}$

$= 8.52 \text{ L}$

25. 2,805,269 acres

$= 2,805,269 \cancel{\text{acres}} \times \dfrac{1 \text{ mi}^2}{640 \cancel{\text{acre}}}$

$= \dfrac{2,805,269}{640} \text{ mi}^2$

$\approx 4,383.23 \text{ mi}^2$

27. $250 \text{ mi} = 250 \ \cancel{\text{mi}} \times \dfrac{1.61 \text{ km}}{1 \ \cancel{\text{mi}}}$

$\qquad\quad = 250 \times 1.61 \text{ km}$

$\qquad\quad = 402.5 \text{ km}$

29. Find the area of 1 brick in m^2 :

$A = (20 \text{ cm})(10 \text{ cm})$

$\quad = 200 \ \cancel{\text{cm}}^2 \times \dfrac{1 \text{ m}}{100 \ \cancel{\text{cm}}} \times \dfrac{1 \text{ m}}{100 \ \cancel{\text{cm}}}$

$\quad = \dfrac{200}{100 \times 100} \text{m}^2$

$\quad = 0.02 \text{ m}^2$

Convert the area of the entryway to bricks:

$12 \text{ m}^2 = 12 \ \cancel{\text{m}^2} \times \dfrac{1 \text{ brick}}{0.02 \ \cancel{\text{m}^2}} = \dfrac{12}{0.02} \text{ bricks}$

$\qquad\quad = 600 \text{ bricks}$

It will take 600 bricks to cover the patio.

31. Convert 5 gal to oz, then glasses:

$5 \text{ gal} = 5 \ \cancel{\text{gal}} \times \dfrac{128 \ \cancel{\text{fl oz}}}{1 \ \cancel{\text{gal}}} \times \dfrac{1 \text{ glass}}{8 \ \cancel{\text{fl oz}}}$

$\qquad\quad = \dfrac{5 \times 128}{8} \text{ glasses}$

$\qquad\quad = 80 \text{ glasses}$

You can fill 80 glasses.

33. $188 \text{ km per hour} = 188 \dfrac{\cancel{\text{km}}}{\text{hr}} \times \dfrac{1 \text{ mi}}{1.61 \ \cancel{\text{km}}}$

$\qquad\qquad\qquad = \dfrac{188}{1.61} \dfrac{\text{mi}}{\text{hr}}$

$\qquad\qquad\qquad \approx 117 \dfrac{\text{mi}}{\text{hr}}$

35. a. $4 \text{ hr } 45 \text{ min} = 4 \ \cancel{\text{hr}} \times \dfrac{60 \text{ min}}{1 \ \cancel{\text{hr}}} + 45 \text{ min}$

$\qquad\qquad\qquad\quad = 240 \text{ min} + 45 \text{ min}$

$\qquad\qquad\qquad\quad = 285 \text{ min}$

  b. $4 \text{ hr } 45 \text{ min} = 4 \text{ hr} + 45 \ \cancel{\text{min}} \times \dfrac{1 \text{ hr}}{60 \ \cancel{\text{min}}}$

$\qquad\qquad\qquad\quad = 4 \text{ hr} + 0.75 \text{ hr}$

$\qquad\qquad\qquad\quad = 4.75 \text{ hr}$

37. Total weight: $\quad$ 12 lb 8 oz

$$\begin{array}{r} \times \qquad 2 \\ \hline 24 \text{ lb } 16 \text{ oz} = 25 \text{ lbs} \end{array}$$

Total cost: $25 \ \cancel{\text{lb}} \times \dfrac{\$5.00}{\cancel{\text{lb}}} = \$125$

Chapters 1-6 Cumulative Review

1.
$$\overset{1\ \ 1}{7,520}$$
$$599$$
$$+\ 8,640$$
$$\overline{16,759}$$

3.
$$13\overline{)156}\ \ \overset{12}{}$$
$$\underline{13\downarrow}$$
$$26$$
$$\underline{26}$$
$$0$$

5. $64\overline{)31,362}\ \overset{490}{} \rightarrow \dfrac{31,362}{64} = 490\dfrac{2}{64} = 490\dfrac{1}{32}$
$$\underline{256\downarrow}$$
$$576$$
$$\underline{576\downarrow}$$
$$02$$

7. $12 + 81 \div 3^2 = 12 + 81 \div 9$
$$= 12 + 9$$
$$= 21$$

9.
$$25$$
$$+\ 13$$
$$\overline{38}$$

11. $\dfrac{39}{3} = \dfrac{\cancel{3} \cdot 13}{\cancel{3}} = 13$

13.
$$5.40$$
$$2.58$$
$$+\ 3.09$$
$$\overline{11.07}$$

15. $25\overline{)40.50}\ \overset{1.62}{}$
$$\underline{25\downarrow}$$
$$155$$
$$\underline{150\downarrow}$$
$$50$$
$$\underline{50}$$
$$0$$

17. $17 \div \left(\dfrac{1}{3}\right)^2 = 17 \div \left(\dfrac{1}{3}\right)\left(\dfrac{1}{3}\right)$
$$= 17 \div \dfrac{1}{9}$$
$$= 17 \cdot \dfrac{9}{1}$$
$$= 153$$

19. $\left(16 \div 1\dfrac{1}{4}\right) \div 2 = \left(\dfrac{16}{1} \div \dfrac{5}{4}\right) \div \dfrac{2}{1}$
$$= \dfrac{16}{1} \cdot \dfrac{4}{5} \cdot \dfrac{1}{2}$$
$$= \dfrac{16 \cdot 2 \cdot \cancel{2} \cdot 1}{1 \cdot 5 \cdot \cancel{2}}$$
$$= \dfrac{32}{5}$$
$$= 6\dfrac{2}{5}$$

21. $\dfrac{3}{8}(2.4)\ \dfrac{3}{5}(0.25) = 0.375(2.4) - 0.6(0.25)$
$$= 0.9 - 0.15$$
$$= 0.75$$

23. $13 + \dfrac{3}{14} \div \dfrac{5}{42} = 13 + \dfrac{3}{14} \cdot \dfrac{42}{5}$

$= 13 + \dfrac{3 \cdot \cancel{14} \cdot 3}{\cancel{14} \cdot 5}$

$= 13 + \dfrac{9}{5}$

$= 13\dfrac{9}{5}$

$= 14\dfrac{4}{5}$

25. $46 = 4 \cdot y$

$\dfrac{46}{4} = \dfrac{\cancel{4} \cdot y}{\cancel{4}}$

$\dfrac{46}{4} = y$

$\dfrac{23}{2} = y$

$11\dfrac{1}{2} = y$

27.

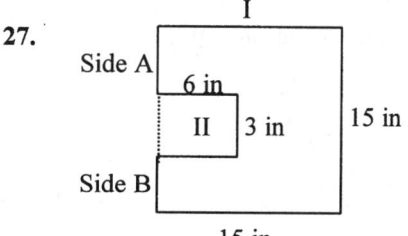

Find the lengths of the missing sides:
Side A + Side B = 15 in − 3 in = 12 in.
Because no other information is given, we will assume Side A and Side B are equal in length, and both are 6 in.

P = sum of all sides

$= 15 + 15 + 15 + 6 + 6 + 3 + 6 + 6$

$= 72$ in

To find the area, let Area I be the large square, and Area II be the small rectangle.

$A = \text{Area I} - \text{Area II}$

$= lw - lw$

$= 15 \cdot 15 - 6 \cdot 3$

$= 225 - 18$

$= 207 \text{ in}^2$

29.
$$\begin{array}{r} 62 \\ -\ 15 \\ \hline 47 \end{array}$$

31. $x = 0.24(7{,}450)$

$x = 1{,}788$

33. 2/3 of the product of 7 and 9

$\quad\quad\downarrow \quad\quad\quad\quad\quad\quad \downarrow$

$\quad\dfrac{2}{3} \times \quad\quad\quad (7 \times 9)$

$\dfrac{2}{3}(7 \cdot 9) = \dfrac{2}{3} \cdot \dfrac{63}{1} = \dfrac{2 \cdot \cancel{3} \cdot 21}{\cancel{3} \cdot 1} = 42$

35. Let x = miles the plane will fly

$$\frac{375 \text{ miles}}{1\frac{1}{2} \text{ hours}} = \frac{x \text{ miles}}{2\frac{3}{4} \text{ hours}}$$

$$375\left(2\frac{3}{4}\right) = 1\frac{1}{2}x$$

$$1031.25 = 1.5x$$

$$\frac{1031.25}{1.5} = \frac{\cancel{1.5}x}{\cancel{1.5}}$$

$$687.5 = x$$

The plane will fly 657.5 miles.

37. $18\frac{1}{3} \cdot 12 = \frac{55}{3} \cdot \frac{12}{1} = \frac{55 \cdot \cancel{3} \cdot 4}{\cancel{3} \cdot 1} = 220 \text{ in.}$

Chapter 6 Test

1. $7 \text{ yd} = 7 \text{ yd} \times \dfrac{3 \text{ ft}}{1 \text{ yd}}$

$= 7 \times 3 \text{ ft}$

$= 21 \text{ ft}$

3. $3 \text{ acres} = 3 \text{ acres} \times \dfrac{43,560 \text{ ft}^2}{1 \text{ acre}}$

$= 3 \times 43,560 \text{ ft}^2$

$= 130,680 \text{ ft}^2$

5. $10 \text{ L} = 10 \text{ L} \times \dfrac{1,000 \text{ mL}}{1 \text{ L}}$

$= 10 \times 1,000 \text{ mL}$

$= 10,000 \text{ mL}$

7. $10 \text{ L} = 10 \text{ L} \times \dfrac{1.06 \text{ qts}}{1 \text{ L}}$

$= 10 \times 1.06 \text{ qt}$

$= 10.6 \text{ qt}$

9. $1 \text{ L} = 1 \text{ L} \times \dfrac{1.06 \text{ qt}}{1 \text{ L}} \times \dfrac{1 \text{ gal}}{4 \text{ qt}}$

$= \dfrac{1 \times 1.06}{4} \text{ gal}$

$= 0.27 \text{ gal}$

11. $409 \text{ in}^3 = 409 \text{ in}^3 \times \dfrac{1 \text{ ft}^3}{1,728 \text{ in}^3}$

$= \dfrac{409}{1,728} \text{ ft}^3$

$= 0.24 \text{ ft}^3$

13. $245 \text{ ft} = 245 \text{ ft} \times \dfrac{1 \text{ m}}{3.28 \text{ ft}}$

$= \dfrac{245}{3.28} \text{ m}$

$= 74.70 \text{ m}$

15. Find the area of 1 tile:

$A = (8 \text{ in})(8 \text{ in}) = 64 \text{ in}^2$

Convert the area of the entryway to tile:

40 ft^2

$= 40 \text{ ft}^2 \times \dfrac{144 \text{ in}^2}{1 \text{ ft}^2} \times \dfrac{1 \text{ tile}}{64 \text{ in}^2}$

$= \dfrac{40 \times 144}{64} \text{ tiles}$

$= 90 \text{ tiles}$

It will take 90 tiles to cover the pantry.

17. a. $5 \text{ hr } 30 \text{ min} = 5 \text{ hr} \times \dfrac{60 \text{ min}}{1 \text{ hr}} + 30 \text{ min}$

$= 300 \text{ min} + 30 \text{ min}$

$= 330 \text{ min}$

b. $5 \text{ hr } 30 \text{ min} = 5 \text{ hr} + 30 \text{ min} \times \dfrac{1 \text{ hr}}{60 \text{ min}}$

$= 5 \text{ hr} + 0.5 \text{ hr}$

$= 5.5 \text{ hr}$

19. a. $80 \text{ km per hour} = 80 \dfrac{\text{km}}{\text{hr}} \times \dfrac{1 \text{ mi}}{1.61 \text{ km}}$

$= \dfrac{80}{1.61} \dfrac{\text{mi}}{\text{hr}}$

$= 50 \dfrac{\text{mi}}{\text{hr}}$

b. You could only out run it with a car.

Chapter 7 Preview

1. Addition facts:

+	0	1	2	3	4	5	6	7	8	9
0	0	1	2	3	4	5	6	7	8	9
1	1	2	3	4	5	6	7	8	9	10
2	2	3	4	5	6	7	8	9	10	11
3	3	4	5	6	7	8	9	10	11	12
4	4	5	6	7	8	9	10	11	12	13
5	5	6	7	8	9	10	11	12	13	14
6	6	7	8	9	10	11	12	13	14	15
7	7	8	9	10	11	12	13	14	15	16
8	8	9	10	11	12	13	14	15	16	17
9	9	10	11	12	13	14	15	16	17	18

3.
$$\begin{array}{r} 60.3 \\ + \ 49.8 \\ \hline 110.1 \end{array}$$

5. $(0.4)(0.8) = 0.32$

7. $5^3 = 5 \cdot 5 \cdot 5 = 125$

9. $14 - 8 = 6$

11. $60 \div 20 = \dfrac{60}{20} = \dfrac{20 \cdot 3}{20} = 3$

13. $\dfrac{7}{8} \cdot \dfrac{5}{14} = \dfrac{7 \cdot 5}{8 \cdot 2 \cdot 7} = \dfrac{5}{16}$

15. $12 \div 4 = 3$

17. $5(3 \cdot 2) = (5 \cdot 3) \cdot 2$

19. $P = 4s$
$\qquad = 4(6 \text{ in})$
$\qquad = 24 \text{ in}$

Chapter 7 Pretest

1. The opposite of 10 is -10.

3. $\dfrac{6}{5} > -\dfrac{5}{6}$, because $\dfrac{6}{5}$ is right of $-\dfrac{5}{6}$ on the number line.

5. $-|-8| = -8$

7. $9 - 19 = 9 + (-19)$
$$= -10$$

9. $-\dfrac{3}{5} - \dfrac{2}{5} = \dfrac{-3 + (-2)}{5}$
$$= -\dfrac{5}{5}$$
$$= -1$$

11. $-40 - (-7) = -40 + 7$
$$= -33$$

13. $-\dfrac{2}{3}\left(-\dfrac{3}{5}\right) = \dfrac{2 \cdot \cancel{3}}{\cancel{3} \cdot 5} = \dfrac{2}{5}$

15. $\dfrac{-40}{-5} = 8$

17. $(-2)^3$ *definition of exponent*
$$= (-2)(-2)(-2) \quad \textit{multiply left to right}$$
$$= 4(-2) \quad\quad\quad \textit{multiply}$$
$$= -8$$

19. Simplify above and below fraction bar first:
$$\dfrac{7 + 4(-1)}{2 - 3} \quad \textit{multiply first}$$
$$= \dfrac{7 + (-4)}{2 + (-3)} \quad \textit{add}$$
$$= \dfrac{3}{-1} \quad\quad\quad \textit{divide}$$
$$= -3$$

21. $5(3y - 2) = 5 \cdot 3y - 5 \cdot 2$
$$= 15y - 10$$

23. High temp $-$ low temp $=$ difference
$$30° - (-5°)$$
$$= 30° + 5°$$
$$= 35°$$

Section 7.1

1. $4 < 7$
   4 is less than 7.

3. $5 > -2$
   5 is greater than -2.

5. $-10 < -3$
   -10 is less than -3.

7. $0 > -4$
   0 is greater than -4.

9. 30 is greater than -30.
   $30 > -30$

11. -10 is less than 0.
    $-10 < 0$

13. -3 is greater than -15.
    $-3 > -15$

15. $3 < 7$, because 3 is left of 7 on the number line.

17. $7 > -5$, because 7 is right of -5 on the number line.

19. $-6 < 0$, because -6 is left of 0 on the number line.

21. $-12 < -2$, because -12 is left of -2 on the number line.

23. $\dfrac{1}{2} > -\dfrac{3}{4}$, because $\dfrac{1}{2}$ is right of $-\dfrac{3}{4}$ on the number line.

25. $-0.75 < -0.25$, because -0.75 is left of 0.25 on the number line.

27. $-0.1 < -0.01$, because -0.1 is left of -0.01 on the number line.

29. $-3\ \underline{\ ?\ }\ |-6|$ replace $|-6| = 6$
    $-3\ \underline{\ ?\ }\ 6$
    $-3 < 6$

31. $15\ \underline{\ ?\ }\ |-4|$ replace $|-4| = 4$
    $15\ \underline{\ ?\ }\ 4$
    $15 > 4$

33. $|-2|\ \underline{\ ?\ }\ |-7|$ replace $|-7| = 7$, $|-2| = 2$
    $2\ \underline{\ ?\ }\ 7$
    $2 < 7$

35. $|2| = 2$, because 2 is 2 units from 0.

37. $|100| = 100$, because 100 is 100 units from 0.

39. $|-8| = 8$, because -8 is 8 units from 0.

41. $|-231| = 231$, because -231 is 231 units from 0.

43. $\left|\dfrac{3}{4}\right| = \dfrac{3}{4}$, because $\dfrac{3}{4}$ is $\dfrac{3}{4}$ units from 0.

45. $|-200| = 200$, because -200 is 200 units from 0.

47. $|8| = 8$, because 8 is 8 units from 0.

49. $|231| = 231$, because 231 is 231 units from 0.

51. The opposite of 3 is -3

53. The opposite of -2 is $-(-2) = 2$.

55. The opposite of 75 is -75.

57. The opposite of 0 is 0.

59. The opposite of -0.123 is $-(-0.123) = 0.123$

61. The opposite of $\frac{7}{8}$ is $-\frac{7}{8}$.

63. $-(-2) = 2$
The opposite of -2 is 2.

65. $-(-8) = 8$
The opposite of -8 is 8.

67. $-|-2| = -2$

69. $-|-8| = -8$

71. The only number that is its own opposite is 0.

73. If n is negative:
$-n$ is $-$(a negative number)
$-n$ is a positive number

Estimating

75. -60 is closer to -100 on the number line.

77. -10 is closer to -20 on the number line.

79. -362 is closer to -360 on the number line.

Applying the Concepts

81. 61 degrees *below zero* Fahrenheit: $-61\,°F$
10 degrees warmer: $-51\,°F$
Note: below zero indicates a negative number.

83. 5 degrees *below zero* Fahrenheit: $-5\,°F$
10 degrees colder: $-15\,°F$
Note: below zero indicates a negative number.

85. From the 25 mph row, 25°F column: $-7\,°F$

87. 10°F and 25 mph wind gives wind chill temperature of $-29°F$.
$-5\,°F$ and 10 mph wind gives wind chill temperature of $-27°F$.
Because $-29 < -27$, 10°F and 25 mph wind feels colder.

89. Above each month (numbered 1-12, beginning with January), draw a dot corresponding with the low temperature as measured by the vertical axis. To sketch the line graph, connect the dots.

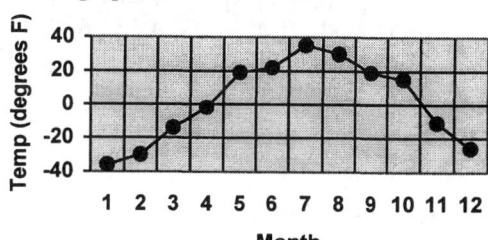

Review Problems

91.   4,005
       276
    $+\ \ 32$
    4,313

93. $\left.\begin{array}{l} 2 = 2 \\ 4 = 2 \cdot 2 \end{array}\right\}$ LCD $= 2 \cdot 2 = 4$

$\dfrac{1}{2} + \dfrac{3}{4} = \dfrac{1 \cdot 2}{2 \cdot 2} + \dfrac{3}{4} = \dfrac{2}{4} + \dfrac{3}{4} = \dfrac{5}{4} = 1\dfrac{1}{4}$

95.   3.780
       1.400
    $+\ 26.701$
    31.881

97.   635
    $-\ 579$
       56

99.
$$\begin{array}{r} 36.24 \\ -\,25.37 \\ \hline 10.87 \end{array}$$

101. $\left.\begin{array}{l} 28 = 2\cdot2\cdot7 \\ 14 = 2\cdot7 \end{array}\right\} \text{LCD} = 2\cdot2\cdot7 = 28$

$$\frac{27}{28} - \frac{11}{14} = \frac{27}{29} - \frac{11\cdot2}{14\cdot2} = \frac{27}{28} - \frac{22}{28} = \frac{5}{28}$$

Extending the Concepts

103.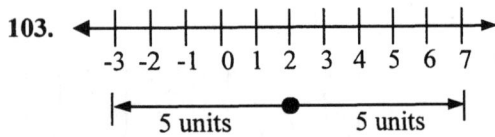

−3 is also 5 units from 2.

105. Answers will vary.

107. Answers will vary.

Improving Your Quantitative Literacy

109. a. The highest wind speed came on the day with the lowest wind chill:
   −59°F wind chill: 1/10/82 Cincinnati

   b. The lowest wind speed cam on the day with the highest wind chill:
   No wind chill: 1/4/81 Cleveland

   c. The highest wind speed has the lowest wind chill, which is at Green Bay ($-48°F$ wind chill).

Section 7.2

1.

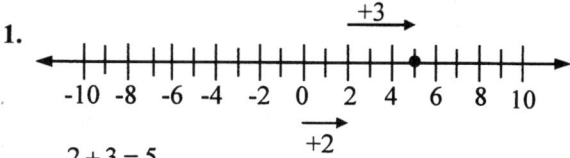

$2 + 3 = 5$

3.

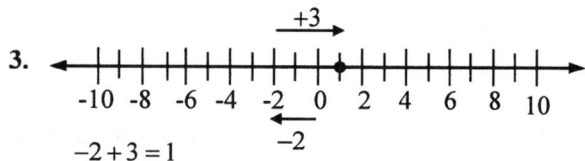

$-2 + 3 = 1$

5.

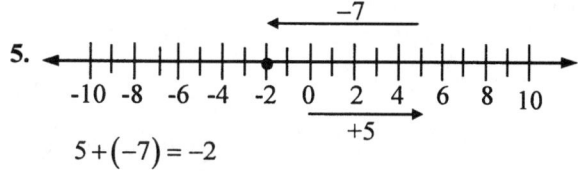

$5 + (-7) = -2$

7.

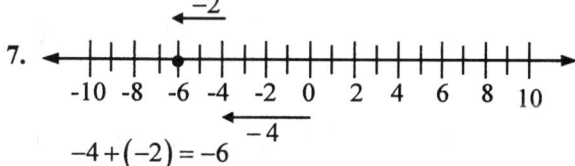

$-4 + (-2) = -6$

9.

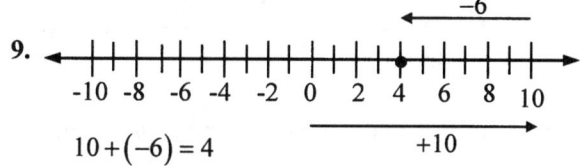

$10 + (-6) = 4$

11.

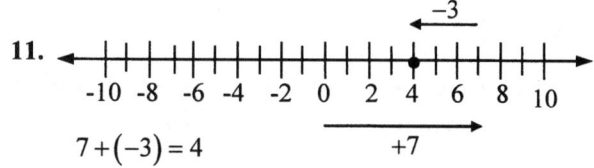

$7 + (-3) = 4$

13.

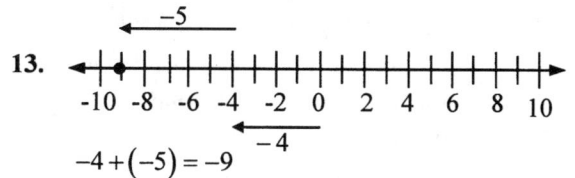

$-4 + (-5) = -9$

15. $7 + 8:$ *add absolute values*

$|7| + |8| = 7 + 8 = 15$

The common sign is positive, so the answer is +15.

17. $5 + (-8):$ *subtract absolute values*

$|-8| - |5| = 8 - 5 = 3$

The number with the larger absolute value is -8, so the answer is -3.

19. $-6 + (-5):$ *add absolute values*

$|-6| + |-5| = 6 + 5 = 11$

The common sign is negative, so the answer is -11.

21. $-10 + 3:$ *subtract absolute values*

$|-10| - |3| = 10 - 3 = 7$

The number with the larger absolute value is -10, so the answer is -7.

23. $-1 + (-2):$ *add absolute values*

$|-1| + |-2| = 1 + 2 = 3$

The common sign is negative, so the answer is -3.

25. $-11 + (-5):$ *add absolute values*

$|-11| + |-5| = 11 + 5 = 16$

The common sign is negative, so the answer is -16.

27. $4 + (-12):$ *subtract absolute values*

$|-12| - |4| = 12 - 4 = 8$

The number with the larger absolute value is -12, so the answer is -8.

29. $-85 + (-42):$ *add absolute values*

$|-85| + |-42| = 85 + 42 = 127$

The common sign is negative, so the answer is -127.

31. $-121+170$: *subtract absolute values*

$|170|-|-121|=170-121=49$

The number with the larger absolute value is
$+170$, so the answer is $+49$.

33. $-375+409$: *subtract absolute values*

$|409|-|-375|=409-375=34$

The number with the larger absolute value is
$+409$, so the answer is $+34$.

35. The table is as follows:

a	b	$a+b$
5	-3	$5+(-3)=2$
5	-4	$5+(-4)=1$
5	-5	$5+(-5)=0$
5	-6	$5+(-6)=-1$
5	-7	$5+(-7)=-2$

37. The table is as follows:

x	y	$x+y$
-5	-3	$-5+(-3)=-8$
-5	-4	$-5+(-4)=-9$
-5	-5	$-5+(-5)=-10$
-5	-6	$-5+(-6)=-11$
-5	-7	$-5+(-7)=-12$

39. $10+(-18)+4$ *add left to right*

$=-8+4$ *add*

$=-4$

41. $24+(-6)+(-8)$ *add left to right*

$=18+(-8)$ *add*

$=10$

43. $-201+(-143)+(-101)$ *add left to right*

$=-344+(-101)$ *add*

$=-445$

45. $-321+752+(-324)$ *add left to right*

$=431+(-324)$ *add*

$=107$

47. $-8+3+(-5)+9$ *add left to right*

$=-5+(-5)+9$ *add left to right*

$=-10+9$ *add*

$=-1$

49. $-2+(-5)+(-6)+(-7)$ *add left to right*

$=-7+(-6)+(-7)$ *add left to right*

$=-13+(-7)$ *add*

$=-20$

51. $15+(-30)+18+(-20)$ *add left to right*

$=-15+18+(-20)$ *add left to right*

$=3+(-20)$ *add*

$=-17$

53. $-78+(-42)+57+13$ *add left to right*

$=-120+57+13$ *add left to right*

$=-63+13$ *add*

$=-50$

55. $(-8+5)+(-6+2)$ *parenthesis first*

$=-3+(-4)$ *add*

$=-7$

57. $(-10+4)+(-3+12)$ *parenthesis first*

$=-6+9$ *add*

$=3$

59. $20+(-30+50)+10$ *parenthesis first*

$=20+20+10$ *add left to right*

$=40+10$ *add*

$=50$

61. $108 + (-456 + 275)$ *parenthesis*

    $= 108 + (-181)$ *add left to right*

    $= -73$

63. $\left[5 + (-8)\right] + \left[3 + (-11)\right]$ *brackets first*

    $= -3 + (-8)$ *add*

    $= -11$

65. $\left[57 + (-35)\right] + \left[19 + (-24)\right]$ *brackets first*

    $= 22 + (-5)$ *add*

    $= 17$

67. Add absolute values. The answer will be negative.

    $-1.3 + (-2.5) = -3.8$

69. Subtract the smaller absolute value from the larger absolute value. 24.8 has the larger absolute value, so the answer will be positive.

    $24.8 + (-10.4) = 14.4$

71. $-5.35 + 2.35 + (-6.89)$ *add left to right*

    $= -3 + (-6.89)$ *add*

    $= -9.89$

73. Add the absolute values. The answer will be negative.

    $-\dfrac{5}{6} + \left(-\dfrac{1}{6}\right) = -\dfrac{6}{6} = -1$

75. Subtract the absolute values. The answer will be negative.

    $\dfrac{3}{7} + \left(-\dfrac{5}{7}\right) = -\dfrac{2}{7}$

77. $-\dfrac{2}{5} + \dfrac{3}{5} + \left(-\dfrac{4}{5}\right)$ *add left to right*

    $= \dfrac{1}{5} + \left(-\dfrac{4}{5}\right)$ *add*

    $= -\dfrac{3}{5}$

79. $-3.8 + 2.54 + 0.4$ *add left to right*

    $= -1.26 + 0.4$ *add*

    $= -0.86$

81. $-2.89 + (-1.4) + 0.09$ *add left to right*

    $= -4.29 + 0.09$ *add*

    $= -4.2$

83. $\dfrac{1}{2} + \left(-\dfrac{3}{4}\right) = \dfrac{1 \cdot 2}{2 \cdot 2} + \left(-\dfrac{3}{4}\right)$

    $= \dfrac{2}{4} + \left(-\dfrac{3}{4}\right)$

    $= -\dfrac{1}{4}$

85. $-\dfrac{2}{3} + \left(-\dfrac{3}{5}\right) + \left(-\dfrac{1}{15}\right)$

    $= -\dfrac{2 \cdot 5}{3 \cdot 5} + \left(-\dfrac{3 \cdot 3}{5 \cdot 3}\right) + \left(-\dfrac{1}{15}\right)$

    $= -\dfrac{10}{15} + \left(-\dfrac{9}{15}\right) + \left(-\dfrac{1}{15}\right)$

    $= -\dfrac{20}{15}$

    $= -1\dfrac{5}{15}$ or $-1\dfrac{1}{3}$

87. $-8 + (-10) + (-3)$ *add left to right*

    $= -18 + (-3)$ *add*

    $= -21$

89.

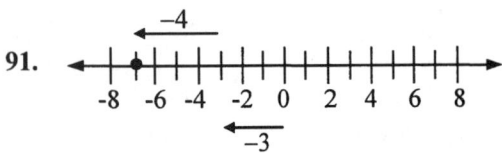

-5, because $8+(-5)=3$

91.

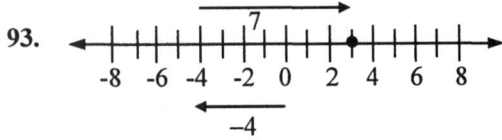

-4, because $-3+(-4)=-7$

93.

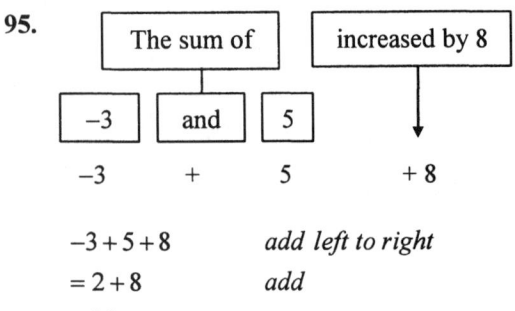

7, because $-4+7=3$

95.

The sum of		increased by 8

-3	and	5

-3	$+$	5	$+8$

$$-3+5+8 \qquad \text{add left to right}$$
$$=2+8 \qquad \text{add}$$
$$=10$$

Estimating

97. a. $251+249$ rounds to $250+250=500$

99. b. $-251+249$ rounds to $-250+250=0$

101. d. $77+22$ rounds to $80+20=100$

103. c. $77+(-22)$ rounds to $80+(-20)=60$

Applying the Concepts

105. Balance + deposit + check = new balance
$$-\$40+\$100+(-\$50)$$
$$=\$60+(-\$50)$$
$$=\$10$$

107. Wins + losses = net loss/gain
$$\$74+(-\$141)=-\$67$$

109. Gains + losses = net loss/gain
$$3 \text{ pts} + (-5 \text{ pts}) = -2 \text{ pts}$$

111.

The answers are -7 and 13, because
$$3+(-10)=-7 \text{ and } 3+10=13$$

Review Problems

113. $7\overset{5\ 13}{\cancel{6}\cancel{3}}$
$$\underline{-1\ 5\ 9}$$
$$6\ 0\ 4$$

115. $465-462-3 \qquad \text{subtract left to right}$
$$=3-3 \qquad \text{subtract}$$
$$=0$$

117. $33.44-22.55-9.89$
$$=10.89-9.89$$
$$=1$$

119. $\dfrac{3}{4} - \dfrac{1}{3} - \dfrac{1}{12} = \dfrac{3 \cdot 3}{4 \cdot 3} - \dfrac{1 \cdot 4}{3 \cdot 4} - \dfrac{1}{12}$

$\qquad\qquad\quad = \dfrac{9}{12} - \dfrac{4}{12} - \dfrac{1}{12}$

$\qquad\qquad\quad = \dfrac{4}{12}$

$\qquad\qquad\quad = \dfrac{1}{3}$

Section 7.3

1. $7-5=7+(-5)=2$

3. $8-6=8+(-6)=2$

5. $-3-5=-3+(-5)=-8$

7. $-4-1=-4+(-1)=-5$

9. $5-(-2)=5+2=7$

11. $3-(-9)=3+9=12$

13. $-4-(-7)=-4+7=3$

15. $-10-(-3)=-10+3=-7$

17. $15-18=15+(-18)=-3$

19. $100-113=100+(-113)=-13$

21. $-30-20=-30+(-20)=-50$

23. $-79-21=-79+(-21)=-100$

25. $156-(-243)=156+243=399$

27. $-35-(-14)=-35+14=-21$

29. The table is as follows:

x	y	$x-y$
8	6	$8-6=8+(-6)=2$
8	7	$8-7=8+(-7)=1$
8	8	$8-8=8+(-8)=0$
8	9	$8-9=8+(-9)=-1$
8	10	$8-10=8+(-10)=-2$

31. The table is as follows:

x	y	$x-y$
8	-6	$8-(-6)=8+6=14$
8	-7	$8-(-7)=8+7=15$
8	-8	$8-(-8)=8+8=16$
8	-9	$8-(-9)=8+9=17$
8	-10	$8-(-10)=8+10=18$

33. $-9.01-2.4=-9.01+(-2.4)=-11.41$

35. $-0.89-1.01=-0.89+(-1.01)=-1.9$

37. $-\dfrac{1}{6}-\dfrac{5}{6}=-\dfrac{1}{6}+\left(-\dfrac{5}{6}\right)=-\dfrac{6}{6}=-1$

39. $\dfrac{5}{12}-\dfrac{5}{6}=\dfrac{5}{12}-\dfrac{5\cdot2}{6\cdot2}$ $LCD=12$

$\qquad =\dfrac{5}{12}+\left(-\dfrac{10}{12}\right)$

$\qquad =-\dfrac{5}{12}$

41. $\left.\begin{array}{l}70=2\cdot5\cdot7\\42=2\cdot3\cdot7\end{array}\right\}$ LCD $=2\cdot3\cdot5\cdot7=210$

$\qquad -\dfrac{13}{70}-\dfrac{23}{42}=-\dfrac{13\cdot3}{70\cdot3}-\dfrac{23\cdot5}{42\cdot5}$

$\qquad\qquad =-\dfrac{39}{210}+\left(-\dfrac{115}{210}\right)$

$\qquad\qquad =-\dfrac{154}{210}$

$\qquad\qquad =-\dfrac{11}{15}$

43. $4-5-6$ *change subtraction to*
  addition of the opposite

$\qquad =4+(-5)+(-6)$ *add from left to right*

$\qquad =-1+(-6)$ *add*

$\qquad =-7$

45. $-8+3-4$ *change subtraction to*
                         addition of the opposite

$= -8+3+(-4)$ *add from left to right*

$= -5+(-4)$ *add*

$= -9$

47. $-8-4-2$ *change subtraction to*
                         addition of the opposite

$= -8+(-4)+(-2)$ *add from left to right*

$= -12+(-2)$ *add*

$= -14$

49. $-3.4-5.6-8.5$ *change subtraction to*
                               addition of the opposite

$= -3.4+(-5.6)+(-8.5)$ *add left to right*

$= -9+(-8.5)$ *add*

$= -17.5$

51. $3-(-2)-6$ *change subtraction to*
                         addition of the opposite

$= 3+2+(-6)$ *add from left to right*

$= 5+(-6)$ *add*

$= -1$

53. $\dfrac{1}{2}-\dfrac{1}{3}-\dfrac{1}{4}=\dfrac{1}{2}+\left(-\dfrac{1}{3}\right)+\left(-\dfrac{1}{4}\right)$

$= \dfrac{1\cdot 6}{2\cdot 6}+\left(-\dfrac{1\cdot 4}{3\cdot 4}\right)+\left(-\dfrac{1\cdot 3}{4\cdot 3}\right)$

$= \dfrac{6}{12}+\left(-\dfrac{4}{12}\right)+\left(-\dfrac{3}{12}\right)$

$= -\dfrac{1}{12}$

55. $-900+400-(-100)$ *change subtraction to*
                             addition of the opposite

$= -900+400+100$ *add from left to right*

$= -500+100$ *add*

$= -400$

57. $5-(-6)=5+6=11$

59. $-5-(-1)=-5+1=-4$

61. $(-8+12)-(-4)$ *change subtration to*
                             addition of the opposite

$= (-8+12)+4$ *add from left to right*

$= 4+4$ *add*

$= 8$

63.

From the number line: $-3+(-6)=-9$
However, adding –6 is the same as
subtracting 6.

Estimating

65. b. $52-49$ rounds to $50-50$

$= 50+(-50)$

$= 0$

67. a. $52-(-49)$ rounds to $50-(-50)$

$= 50+50$

$= 100$

69. $-161-(-62)$ rounds to $-150-(-50)$

$= -150+50$

$= -100$

71. b. $37-61$ rounds to $40-60$

$= 40+(-60)$

$= -20$

73. a. $-37-61$ rounds to $-40-60$

$= -40+(-60)$

$= -100$

Applying the Concepts

75. High temp − low temp = difference
$$28° - (-16°)$$
$$= 28° + 16°$$
$$= 44°$$

77. 2008 price − 2000 price = difference
$$\$1,903 - \$9,809 = \$1,903 + (-\$9,809)$$
$$= -\$7,906$$

79. To find the difference, subtract the 25 mph wind chill temp from the 15 mph wind chill temp:
$$-11°F - (-22°F)$$
$$= -11°F + 22°F$$
$$= 11°F$$

81. To find the difference, subtract the 20 mph wind chill temp from the 10 mph wind chill temp:
$$3°F - (-24°F)$$
$$= 3°F + 24°F$$
$$= 27°F$$

83. To find the difference, subtract the low temperature from the high temperature:
$$60°F - (-26°F)$$
$$= 60°F + 26°F$$
$$= 86°F$$

85. To find the difference:
March low temp − December low temp
$$-14°F - (-26°F)$$
$$= -14°F + 26°F$$
$$= 12°F$$

Review Problems

87.
$$\begin{array}{r} 42 \\ \times\ 56 \\ \hline 252 \\ 2{,}100 \\ \hline 2{,}352 \end{array}$$

89.
$$\begin{array}{r} 3.4 \\ \times\ 0.05 \\ \hline 0.170 \text{ or } 0.17 \end{array}$$

91. $\dfrac{3}{4}\left(\dfrac{2}{3}\right) = \dfrac{\cancel{3}\cdot\cancel{2}}{\cancel{2}\cdot 2 \cdot \cancel{3}} = \dfrac{1}{2}$

93. $\begin{aligned} &2 + 3(4+1) && \textit{parenthesis} \\ &= 2 + 3(5) && \textit{multiply} \\ &= 2 + 15 && \textit{add} \\ &= 17 \end{aligned}$

95. $\begin{aligned} &(6+2)(6-2) && \textit{parenthesis} \\ &= (8)(4) && \textit{multiply} \\ &= 32 \end{aligned}$

97. $5^2 = 5 \cdot 5 = 25$

99. $2^3 \cdot 3^2 = 2 \cdot 2 \cdot 2 \cdot 3 \cdot 3 = 72$

Extending the Concepts

101. Choose any two different integers and find the values of $a-b$ and $b-a$. For example,
$$2 - 6 = 2 + (-6) \qquad 6 - 2 = 6 + (-2)$$
$$= -4 \qquad\qquad\qquad = 4$$
Because $-4 \neq 4$, $2-6 \neq 6-2$. Therefore, subtraction is not commutative.

103. $\$10 - \$12 = -\$2$

A checkbook balance is $10, and you write a check for $12. The new balance is –$2.

105. 10, 5, 0,…is an arithmetic sequence starting with 10. Each number thereafter comes from adding –5 to the previous number. The next numbers in the sequence are –5, –10.

107. –10, –6, –2,…is an arithmetic sequence starting with –10. Each number thereafter comes from adding 4 to the previous number. The next numbers in the sequence are 2, 6.

Improving Your Quantitative Literacy

109. a. 2003 grants – 1998 grants = difference
   $$\$2,484 - \$1,940 = \$544$$

   b. 2003 costs – 1998 costs = difference
   $$\$1,115 - \$1,636 = \$1,115 + (-\$1,636)$$
   $$= -\$521$$

   c. Students are receiving more grants and tax breaks.

Section 7.4

1. $7(-8) = -56$
unlike signs; negative answer

3. $-6(10) = -60$
unlike signs; negative answer

5. $-7(-8) = 56$
like signs; positive answer

7. $-9(-9) = 81$
like signs; positive answer

9. $-2.1(4.3) = -9.03$
unlike signs; negative answer

11. $-\dfrac{4}{5}\left(-\dfrac{15}{28}\right) = \dfrac{4 \cdot 3 \cdot 5}{5 \cdot 4 \cdot 7} = \dfrac{3}{7}$
like signs; positive answer

13. $-12\left(\dfrac{2}{3}\right) = -\dfrac{3 \cdot 4 \cdot 2}{3} = -8$
unlike signs; negative answer

15. $3(-2)(4)$ *multiply left to right*
$= -6(4)$ *multiply*
$= -24$

17. $-4(3)(-2)$ *multiply left to right*
$= -12(-2)$ *multiply*
$= 24$

19. $-1(-2)(-3)$ *multiply left to right*
$= 2(-3)$ *multiply*
$= -6$

21. a. $(-4)^2 = (-4)(-4)$
              $= 16$
   b. $-4^2 = -(4 \cdot 4) = -16$

23. a. $(-5)^3$ *definition of exponent*
      $= (-5)(-5)(-5)$ *multiply left to right*
      $= 25(-5)$ *multiply*
      $= -125$
   b. $-5^3 = -(5 \cdot 5 \cdot 5) = -125$

25. a. $(-2)^4$ *exponent*
      $= (-2)(-2)(-2)(-2)$ *multiply left to right*
      $= 4(-2)(-2)$ *multiply left to right*
      $= -8(-2)$ *multiply*
      $= 16$
   b. $-2^4 = -(2 \cdot 2 \cdot 2 \cdot 2) = -16$

27. The table is as follows:

x	x^2
-3	$(-3)^2 = (-3)(-3) = 9$
-2	$(-2)^2 = (-2)(-2) = 4$
-1	$(-1)^2 = (-1)(-1) = 1$
0	$0^2 = (0)(0) = 0$
1	$1^2 = (1)(1) = 1$
2	$2^2 = (2)(2) = 4$
3	$3^2 = (3)(3) = 9$

29. The table is as follows:

x	y	xy
6	2	$6(2) = 12$
6	1	$6(1) = 6$
6	0	$6(0) = 0$
6	-1	$6(-1) = -6$
6	-2	$6(-2) = -12$

31. The table is as follows:

a	b	ab
-5	3	$(-5)(3) = -15$
-5	2	$(-5)(2) = -10$
-5	1	$(-5)(1) = -5$
-5	0	$(-5)(0) = 0$
-5	-1	$(-5)(-1) = 5$
-5	-2	$(-5)(-2) = 10$
-5	-3	$(-5)(-3) = 15$

33. $4(-3+2)$ *parenthesis first*
$= 4(-1)$ *multiply*
$= -4$

35. $-10(-2-3)$ *parenthesis first*
$= -10(-5)$ *multiply*
$= 50$

37. $-3+2(5-3)$ *parenthesis first*
$= -3+2(2)$ *multiply*
$= -3+4$ *add*
$= 1$

39. $-7+2[-5-9]$ *brackets first*
$= -7+2[-14]$ *multiply*
$= -7+(-28)$ *add*
$= -35$

41. $2(-5)+3(-4)$ *multiply first*
$= -10+(-12)$ *add*
$= -22$

43. $3(-2)4+3(-2)$ *multiply first*
$= -6(4)+(-6)$ *multiply*
$= -24+(-6)$ *add*
$= -30$

45. $(8-3)(2-7)$ *parenthesis first*
$= 5(-5)$ *multiply*
$= -25$

47. $(2-5)(3-6)$ *parenthesis first*
$= -3(-3)$ *multiply*
$= 9$

49. $3(5-8)+4(6-7)$ *parenthesis first*
$= 3(-3)+4(-1)$ *multiply*
$= -9+(-4)$ *add*
$= -13$

51. $-3(4-7)-2(-3-2)$ *parenthesis first*
$= -3(-3)-2(-5)$ *multiply left to right*
$= 9-(-10)$ *change to addition*
$= 9+10$ *add*
$= 19$

53. $3(-2)(6-7)$ *parenthesis first*
$= 3(-2)(-1)$ *multiply left to right*
$= -6(-1)$ *multiply*
$= 6$

55. $-3(-2)(-1)$ *multiply left to right*
$= 6(-1)$ *multiply*
$= -6$

57. $-3(?) = 12$
The answer is -4, because $-3(-4) = 12$

59. $-5(4)-(-3)$ *multiply first*
$= -20-(-3)$ *change to addition*
$= -20+3$ *add*
$= -17$

Applying the Concepts

61. Number of shares × $gain/loss per share = total $gain/loss

$$(100 \text{ shares})(-\$2)+(50 \text{ shares})(\$8)$$
$$= -\$200+\$400$$
$$= \$200$$

63. Set up a table showing a −4°F temperature change for each 1000 ft altitude change:

Altitude Change	Temperature Change
2000 ft to 3000 ft	−4°F
3000 ft to 4000 ft	−4°F
4000 ft to 5000 ft	−4°F
5000 ft to 6000 ft	−4°F
Net Temperature Change	−16°F

65. $\text{Pts} = W(2)+L(-2)+S(3)+TS(4)+BS(-2)$

Gagne:
$$\text{Pts} = 2(2)+3(-2)+53(3)+2(4)+0(-2)$$
$$= 4+(-6)+159+8+0$$
$$= 165$$

Smoltz:
$$\text{Pts} = 0(2)+2(-2)+42(3)+3(4)+4(-2)$$
$$= 0+(-4)+126+12+(-8)$$
$$= 126$$

Wagner:
$$\text{Pts} = 1(2)+4(-2)+40(3)+4(4)+3(-2)$$
$$= 2+(-8)+120+16+(-6)$$
$$= 124$$

Worrell:
$$\text{Pts} = 4(2)+4(-2)+36(3)+2(4)+7(-2)$$
$$= 8+(-8)+108+8+(-14)$$
$$= 102$$

Borowksi:
$$\text{Pts} = 2(2)+2(-2)+32(3)+2(4)+4(-2)$$
$$= 4+(-4)+96+8+(-8)$$
$$= 96$$

67. The table is as follows:

	Value	Number	Product
Eagle	−2	0	$-2(0)=0$
Birdie	−1	7	$-1(7)=-7$
Par	0	7	$0(7)=0$
Bogie	+1	3	$1(3)=3$
Double Bogie	+2	1	$2(1)=2$
			Total: −2

The total of −2 means Garcia was 2 under par. If par was 72, his score was $72+(-2)=70$.

Estimating

69. $-32(-522)$ rounds to $-30(-500) = 15,000$

71. $-47(470)$ rounds to $-50(500) = -25,000$

73. $-222(-987)$ rounds to
$-200(-1,000) = 200,000$

75. $-222 - (-987)$ rounds to
$$-200 - (-1,000) = -200 + 1,000$$
$$= 800$$

Review Problems

77.

```
        The quotient of
   ┌──────┬──────┬──────┐
   │  12  │ and  │  6   │
   └──────┴──────┴──────┘
     12      ÷       6
```

79. $2(3) = 6 \Leftrightarrow 6 \div 3 = 2$

81. $10 \div 5 = 2 \Leftrightarrow 5 \cdot 2 = 10$

83.
$$\begin{array}{r}
89 \\
56\overline{)4,984} \\
448\downarrow \\
\hline
504 \\
504 \\
\hline
0
\end{array}$$

Extending the Concepts

85. a. True. The sum of two negative numbers is always a negative number.

b. False. The statement does not indicate if it is about addition or multiplication. From 85a, we know the *sum* of two negatives is negative.

c. True. The product of two negative numbers is always a positive number.

87. You own 3 shares of a stock which loses $4 per share on a trading day. Your net loss is $3(-\$4) = -\$12$.

89. $2, -6, 18,\ldots$
The geometric sequence starts with 2. Note that $2(-3) = -6$ and $-6(-3) = 18$. Each number after 2 comes from multiplying the previous number by -3. The next numbers in the sequence are -54 and 162.

91. $-2, 6, -18, \ldots$
The geometric sequence starts with -2. Note that $-2(-3) = 6$ and $6(-3) = -18$.

Each number after -2 comes from multiplying the previous number by -3. The next numbers in the sequence are 54 and -162.

93. $5(-2)^2 - 3(-2)^3$ *exponents first*
$$= 5(4) - 3(-8) \quad \text{\textit{multiply}}$$
$$= 20 - (-24) \quad \text{\textit{change to addition}}$$
$$= 20 + 24 \quad \text{\textit{add}}$$
$$= 44$$

95. $7 - 3(4 - 8)$ *inside parenthesis first*

$= 7 - 3(-4)$ *multiply*

$= 7 - (-12)$ *change to addition*

$= 7 + 12$ *add*

$= 19$

97. $5 - 2\left[3 - 4(6 - 8)\right]$ *grouping symbols first*

$= 5 - 2\left[3 - 4(-2)\right]$ *multiply inside brackets*

$= 5 - 2\left[3 - (-8)\right]$ *brackets, change to add*

$= 5 - 2\left[3 + 8\right]$ *inside brackets*

$= 5 - 2\left[11\right]$ *multiply*

$= 5 - 22$ *subtract*

$= -17$

Section 7.5

1. $-15 \div 5 = -3$ *unlike signs, negative answer*

3. $20 \div (-4) = -5$ *unlike signs, negative answer*

5. $-30 \div (-10) = 3$ *like signs, positive answer*

7. $\dfrac{-14}{-7} = 2$ *like signs, positive answer*

9. $\dfrac{12}{-3} = -4$ *unlike signs, negative answer*

11. $\dfrac{-22}{11} = -2$ *unlike signs, negative answer*

13. $\dfrac{0}{-3} = 0$, because $-3(0) = 0$

15. $\dfrac{125}{-25} = -5$ *unlike signs, negative answer*

17. The table is as follows:

a	b	$\dfrac{a}{b}$
100	−5	$\dfrac{100}{-5} = -20$
100	−10	$\dfrac{100}{-10} = -10$
100	−25	$\dfrac{100}{-25} = -4$
100	−50	$\dfrac{100}{-50} = -2$

19. The table is as follows:

a	b	$\dfrac{a}{b}$
−100	−5	$\dfrac{-100}{-5} = 20$
−100	5	$\dfrac{-100}{5} = -20$
100	−5	$\dfrac{100}{-5} = -20$
100	5	$\dfrac{100}{5} = 20$

21. $\dfrac{4(-7)}{-28}$ *simplify above fraction bar first*

$= \dfrac{-28}{-28}$ *divide*

$= 1$

23. $\dfrac{-3(-10)}{-5}$ *simplify above fraction bar first*

$= \dfrac{30}{-5}$ *divide*

$= -6$

25. Simplify above and below fraction bar first:

$\dfrac{2(-3)}{6-3}$

$= \dfrac{-6}{3}$ *divide*

$= -2$

27. Simplify above and below fraction bar first:

$\dfrac{4-8}{8-4}$ *subtract*

$= \dfrac{-4}{4}$ *divide*

$= -1$

29. Simplify above and below fraction bar first:

$\dfrac{2(-3)+10}{-4}$ *multiply before adding*

$=\dfrac{-6+10}{-4}$ *add*

$=\dfrac{4}{-4}$ *divide*

$=-1$

31. Simplify above and below fraction bar first:

$\dfrac{2+3(-6)}{4-12}$ *multiply before adding*

$=\dfrac{2+(-18)}{4+(-12)}$ *change subtraction to addition*

$=\dfrac{-16}{-8}$ *divide*

$=2$

33. Simplify above and below fraction bar first:

$\dfrac{6(-7)+3(-2)}{20-4}$ *multiply before adding*

$=\dfrac{-42+(-6)}{20+(-4)}$ *change subtraction to addition*

$=\dfrac{-48}{16}$ *divide*

$=-3$

35. Simplify above and below fraction bar before dividing:

$\dfrac{3(-7)(-4)}{6(-2)}$ *multiply from left to right*

$=\dfrac{84}{-12}$ *divide*

$=-7$

37. $(-5)^2+20\div4$ *exponents first*

$=25+20\div4$ *divide before adding*

$=25+5$ *add*

$=30$

39. $100\div(-5)^2$ *exponents before division*

$=100\div25$ *divide*

$=4$

41. $-100\div10\div2$ *divide from left to right*

$=-10\div2$ *divide*

$=-5$

43. $-100\div(10\div2)$ *inside parenthesis first*

$=-100\div5$ *divide*

$=-20$

45. $(-100\div10)\div2$ *inside parenthesis first*

$=-10\div2$ *divide*

$=-5$

47. The quotient of -25 and 5:

$\dfrac{-25}{5}=-5$

49. If n represents the number,

$\dfrac{n}{-5}=-7$

$n=35$, because $\dfrac{35}{-5}=-7$

51. Subtract -3 from the quotient of 27 and 9:

$\dfrac{27}{9}-(-3)$ *divide before subtracting*

$=3-(-3)$ *change subtraction to addition*

$=3+3$ *add*

$=6$

Estimating

53. $397\div(-401)$ rounds to $400\div(-400)=-1$

55. $-121\div27$ rounds to $-120\div30=-4$
Note: round to numbers that are easily divisible

57. $-151 + (-40)$ rounds to
$-150 + (-50) = -200$

59. $-151(-49)$ rounds to $-150(-50) = 7,500$

Applying the Concepts

61. Mean = $\dfrac{\text{sum of values}}{\text{number of values}}$

$= \dfrac{-5 + 0 + 5}{3}$

$= \dfrac{0}{3} = 0$

63. a. Mean = $\dfrac{\text{sum of values}}{\text{number of values}}$

$= \dfrac{-5 + (-4) + 0 + 2 + 2 + 2 + 3}{7}$

$= \dfrac{0}{7}$

$= 0$

b. $-5, -4, 0, 2, 2, 2, 3$
The values are already ordered. Because there is an odd number of values, the median is the middle value
Median = 2

c. $-5, -4, 0, 2, 2, 2, 3$
The mode is the value that occurs most often. Mode = 2.

65. Above each day, draw a dot corresponding with the low temperature as measured by the vertical axis. To sketch the line graph, connect the dots.

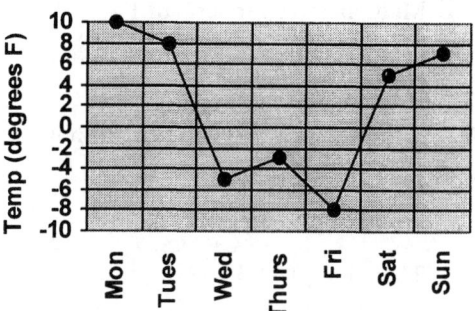

67. a. Mean = $\dfrac{\text{sum of values}}{\text{number of values}}$

$= \dfrac{10 + 8 + (-5) + (-3) + (-8) + 5 + 7}{7}$

$= \dfrac{14}{7}$

$= 2°F$

The mean low temperature is 2°F.

b. Order the values from smallest to largest:
$-8°F, -5°F, -3°F, 5°F, 7°F, 8°F, 10°F$
With an odd number of values, the median is the middle value:
Median = 5°F

Review Problems

69. By the Commutative Property of Addition,
$3 + x = x + 3$

71. By the Associative Property of Addition,
$5 + (7 + a) = (5 + 7) + a$

73. By the Associative Property of Multiplication, $3(4y) = (3 \cdot 4)y$

75. By the Distributive Property,
$5(3 + 7) = 5(3) + 5(7)$

77. $a + 0 = a$

Additive Property of 0

79. $1 \cdot x = x$

Multiplicative Property of 1

Extending the Concepts

81. Choose 2 different numbers to show

$a \div b \neq b \div a$

For example, $3 \div 1 = 3$, but $1 \div 3 = \dfrac{1}{3}$

Therefore, $3 \div 1 \neq 1 \div 3$.

83. Answers will vary.

85. $32, -16, 8, \ldots$

The first number in the sequence is 32.

Note that $32 \div (-2) = -16$ and

$-16 \div (-2) = 8$, i.e., each number after 32 in

the sequence comes from *dividing* the previous number by -2. The next number in the sequence is $8 \div (-2) = -4$.

87. $-32, 16, -8, \ldots$

The first number in the sequence is -32.

Note that $-32 \div (-2) = 16$ and

$16 \div (-2) = -8$, i.e., each number after -32 in the sequence comes from *dividing* the previous number by -2. The next number in the sequence is $-8 \div (-2) = 4$.

89. Simplify above and below fraction bar before dividing:

$\dfrac{-5 - 3}{-5 + 3}$ *change subtraction to addition*

$= \dfrac{-5 + (-3)}{-5 + 3}$ *add*

$= \dfrac{-8}{-2}$ *divide*

$= 4$

91. Simplify above and below fraction bar before dividing:

$\dfrac{6 - 3(2 - 11)}{6 - 3(2 + 11)}$ *inside parenthesis first*

$= \dfrac{6 - 3(-9)}{6 - 3(13)}$ *multiply before adding*

$= \dfrac{6 - (-27)}{6 - 39}$ *change to addition*

$= \dfrac{6 + 27}{6 + (-39)}$ *add*

$= \dfrac{33}{-33}$ *divide*

$= -1$

93. Simplify above and below fraction bar before dividing:

$\dfrac{6 - (3 - 4) - 3}{1 - 2 - 3}$ *inside parenthesis first*

$= \dfrac{6 - (-1) - 3}{1 - 2 - 3}$ *subtract from left to right*

$= \dfrac{4}{-4}$ *divide*

$= -1$

Section 7.6

1. $5(4a) = (5 \cdot 4) a$
$\qquad = 20a$

3. $6(8a) = (6 \cdot 8) a$
$\qquad = 48a$

5. $-6(3x) = (-6 \cdot 3) x$
$\qquad = -18x$

7. $-3(9x) = (-3 \cdot 9) x$
$\qquad = -27x$

9. $5(-2y) = (5 \cdot -2) y$
$\qquad = -10y$

11. $6(-10y) = (6 \cdot -10) y$
$\qquad = -60y$

13. $2 + (3 + x) = (2 + 3) + x$
$\qquad = 5 + x$

15. $5 + (8 + x) = (5 + 8) + x$
$\qquad = 13 + x$

17. $4 + (6 + y) = (4 + 6) + y$
$\qquad = 10 + y$

19. $7 + (1 + y) = (7 + 1) + y$
$\qquad = 8 + y$

21. $(5x + 2) + 4 = 5x + (2 + 4)$
$\qquad = 5x + 6$

23. $(6y + 4) + 3 = 6y + (4 + 3)$
$\qquad = 6y + 7$

25. $(12a + 2) + 19 = 12a + (2 + 19)$
$\qquad = 12a + 21$

27. $(7x + 8) + 20 = 7x + (8 + 20)$
$\qquad = 7x + 28$

29. $7(x + 5) = 7 \cdot x + 7 \cdot 5$
$\qquad = 7x + 35$

31. $6(a - 7) = 6 \cdot a - 6 \cdot 7$
$\qquad = 6a - 42$

33. $2(x - y) = 2 \cdot x - 2 \cdot y$
$\qquad = 2x - 2y$

35. $4(5 + x) = 4 \cdot 5 + 4 \cdot x$
$\qquad = 20 + 4x$

37. $3(2x + 5) = 3 \cdot 2x + 3 \cdot 5$
$\qquad = 6x + 15$

39. $6(3a + 1) = 6 \cdot 3a + 6 \cdot 1$
$\qquad = 18a + 6$

41. $2(6x - 3y) = 2 \cdot 6x - 2 \cdot 3y$
$\qquad = 12x - 6y$

43. $5(7 - 4y) = 5 \cdot 7 - 5 \cdot 4y$
$\qquad = 35 - 20y$

45. $3x + 5x = (3 + 5) x$
$\qquad = 8x$

47. $3a + a = 3a + 1a$
$\qquad = (3 + 1) a$
$\qquad = 4a$

49. $-2x + 6x = (-2 + 6) x$
$\qquad = 4x$

51. $6y - y = 6y - 1y$

$\qquad = (6-1)y$

$\qquad = 5y$

53. $-8a - 2a = (-8-2)a$

$\qquad = -10a$

55. $4x - 9x = (4-9)x$

$\qquad = -5x$

Applying the Concepts

57. $A = s^2 \qquad\qquad P = 4s$

$\quad = (6\text{ ft})^2 \qquad\quad = 4(6\text{ ft})$

$\quad = 36\text{ ft}^2 \qquad\quad = 24\text{ ft}$

59. $A = s^2 \qquad\qquad P = 4s$

$\quad = (9\text{ in})^2 \qquad\quad = 4(9\text{ in})$

$\quad = 81\text{ in}^2 \qquad\quad = 36\text{ in}$

61. $A = lw \qquad\qquad P = 2l + 2w$

$\quad = (20\text{ in})(10\text{ in}) \quad = 2(20\text{ in}) + 2(10\text{ in})$

$\quad = 200\text{ in}^2 \qquad\qquad = 40\text{ in} + 20\text{ in}$

$\qquad\qquad\qquad\qquad\quad = 60\text{ in}$

63. $A = lw \qquad\qquad P = 2l + 2w$

$\quad = (25\text{ ft})(12\text{ ft}) \quad = 2(25\text{ ft}) + 2(12\text{ ft})$

$\quad = 300\text{ ft}^2 \qquad\qquad = 50\text{ ft} + 24\text{ ft}$

$\qquad\qquad\qquad\qquad\quad = 74\text{ ft}$

65. $\dfrac{5(F-32)}{9}$ *substitute value for F*

$= \dfrac{5(68-32)}{9}$ *parenthesis first*

$= \dfrac{5 \cdot 36}{9}$ *divide before multiplying*

$= 5 \cdot \dfrac{36}{9}$ *divide*

$= 5 \cdot 4$ *multiply*

$= 20$

68°F converts to 20°C.

67. $\dfrac{5(F-32)}{9}$ *substitute value for F*

$= \dfrac{5(41-32)}{9}$ *inside parenthesis first*

$= \dfrac{5 \cdot 9}{9}$ *divide before multiplying*

$= 5 \cdot \dfrac{9}{9}$ *divide*

$= 5 \cdot 1$ *multiply*

$= 5$

41°F converts to 5°C.

69. $\dfrac{5(F-32)}{9}$ *substitute value for F*

$= \dfrac{5(14-32)}{9}$ *inside parenthesis first*

$= \dfrac{5 \cdot (-18)}{9}$ *divide before multiplying*

$= 5 \cdot \dfrac{-18}{9}$ *divide*

$= 5 \cdot -2$ *multiply*

$= -10$

14°F converts to −10°C.

Review Problems

71. $8 + (-4) = 4$

73. $-8 + (-4) = -12$

75. $8 - (-4) = 8 + 4$
$\qquad\qquad = 12$

77. $8(-4) = -32$

79. $-8(-4) = 32$

81. $-8 \div 4 = -2$

Chapter 7 Review

1. The opposite of 17 is -17.

3. The opposite of -4.6 is $-(-4.6) = 4.6$

5. $6 > -6$, because 6 is right of -6 on the number line.

7. $|-3| \underline{\ ?\ } 2 \qquad$ *replace* $|-3| = 3$
$\ 3 \ \underline{?\ \ } 2$
$\ 3 \ > \ 2$

9. $-(-4) = 4$
The opposite of -4 is 4.

11. $|-6| = 6$, because -6 is 6 units from 0.

13. $5 + (-7): \qquad$ *subtract absolute values*
$|-7| - |5| = 7 - 5 = 2$
The number with the larger absolute value is negative, so the answer is -2.

15. $-345 + (-626): \quad$ *add absolute values*
$|-345| + |-626| = 345 + 626 = 971$
The common sign is negative, so the answer is -971.

17. $7 - 9 - 4 - 6 \qquad\qquad$ *change to addition*
$= 7 + (-9) + (-4) + (-6) \quad$ *add left to right*
$= -2 + (-4) + (-6) \qquad$ *add left to right*
$= -6 + (-6) \qquad\qquad$ *add*
$= -12$

19. $4 - (-3) \quad$ *change to addition*
$= 4 + 3 \quad$ *add*
$= 7$

21. $5(-4) = -20$
unlike signs; negative answer

23. $(5.6)(-3.1) = -17.36$
unlike signs; negative answer

25. $\dfrac{48}{-16} = -3$
unlike signs; negative answer

27. $\dfrac{-14}{-7} = 2$
like signs; positive answer

29. $(-6)^2 = (-6)(-6) = 36$

31. $(-2)^3 = (-2)(-2)(-2) = -8$

33. $7 + 4(6 - 9) \qquad$ *parenthesis first*
$= 7 + 4(-3) \qquad$ *multiply*
$= 7 + (-12) \qquad$ *add*
$= -5$

35. $(7 - 3)(7 - 9) \qquad$ *parenthesis first*
$= (4)(-2) \qquad\quad$ *multiply*
$= -8$

37. $\dfrac{8 - 4}{-8 + 4} \qquad$ *simplify above and below*
$\qquad\qquad$ *fraction bar first*
$= \dfrac{4}{-4} \qquad$ *divide*
$= -1$

39. $\dfrac{8(-2) + 5(-4)}{12 - 3} \qquad$ *simplify above and below*
$\qquad\qquad\qquad$ *fraction bar first*
$= \dfrac{-16 + (-20)}{9} \qquad$ *add*
$= \dfrac{-36}{9} \qquad$ *divide*
$= -4$

41. $-19+(-23)=-42$

43. $-6-5=-6+(-5)$
$\qquad =-11$

45. $-9(3)=-27$

47. $\dfrac{8(-4)}{-16}=\dfrac{-32}{-16}=2$

49. False, because $\dfrac{-10}{-5}=2$

51. True. $2(-3)=-6$ and $-3+(-3)=-6$.
Therefore, $2(-3)=-3+(-3)$

53. False.
$3-5=-2$ and $5-3=2$.
Therefore, $3-5\neq 5-3$.
Note: Subtraction is not commutative.

55. Wins + losses = net loss/gain
$\$58+(-\$86)=-\$28$

57. High temp – low temp = difference
$17°-(-7°)$
$=17°+7°$
$=24°$

59. $(3x+4)+8=3x+(4+8)$
$\qquad\qquad\quad =3x+12$

61. $-3(7a)=(-3\cdot 7)a$
$\qquad\quad =-21a$

63. $4(x+3)=4\cdot x+4\cdot 3$
$\qquad\quad =4x+12$

65. $7(3y-8)=7(3y)-7\cdot 8$
$\qquad\qquad =(7\cdot 3)y-56$
$\qquad\qquad =21y-56$

67. $7x-4x=(7-4)x=3x$

69. $5y-y=(5-1)y=4y$

Chapters 1-7 Cumulative Review

1. $LCD = 5 \cdot 7 = 35$

$$\frac{3}{5} + \frac{2}{7} = \frac{3 \cdot 7}{5 \cdot 7} + \frac{2 \cdot 5}{7 \cdot 5}$$

$$= \frac{21}{35} + \frac{10}{35}$$

$$= \frac{31}{35}$$

3.
$$\begin{array}{r} {}^{5}\cancel{6}{}^{10}\cancel{1}{}^{13}3 \\ -2\,97 \\ \hline 3\,16 \end{array}$$

5. $\left(\dfrac{2}{3}\right)^4 = \dfrac{2}{3} \cdot \dfrac{2}{3} \cdot \dfrac{2}{3} \cdot \dfrac{2}{3} = \dfrac{16}{81}$

7. $\dfrac{5}{8} \div (-10) = \dfrac{5}{8} \cdot \left(-\dfrac{1}{10}\right) = -\dfrac{\cancel{5} \cdot 1}{8 \cdot 2 \cdot \cancel{5}} = -\dfrac{1}{16}$

9. $4\dfrac{7}{8} = \dfrac{8 \cdot 4 + 7}{8} = \dfrac{39}{8}$

11. $5(x+9) = 5(x) + 5(9)$

Distributive Property

13. $\sqrt{\dfrac{36}{49}} = \dfrac{6}{7}$, because $\left(\dfrac{6}{7}\right)^2 = \dfrac{36}{49}$

15. $(0.2)^3 + (0.3)^2$

$= (0.2)(0.2)(0.2) + (0.3)(0.3)$

$= 0.008 + 0.09$

$= 0.098$

17. $-(-6) = 6$

The opposite of -6 is 6.

19. $\dfrac{\frac{2}{3}}{\frac{3}{4}} = \dfrac{2}{3} \cdot \dfrac{4}{3} = \dfrac{2 \cdot 4}{3 \cdot 3} = \dfrac{8}{9}$

21. $4\dfrac{7}{8} = \dfrac{39}{8} \Rightarrow 8\overline{)39.000}$

$$\begin{array}{r} 4.875 \\ 8\overline{)39.000} \\ 32\downarrow \\ \hline 70 \\ 64\downarrow \\ \hline 60 \\ 56\downarrow \\ \hline 40 \\ 40 \\ \hline 0 \end{array}$$

23. $76\% = \dfrac{76}{100} = \dfrac{19}{25}$

25. 17 is what percent of 42.5?

$17 = n \cdot 42.5$

$\dfrac{17}{42.5} = \dfrac{n \cdot \cancel{42.5}}{\cancel{42.5}}$

$n = \dfrac{17}{42.5}$

$n = 0.4 = 40\%$

27. $14 \text{ gal} = 14 \ \cancel{\text{gal}} \times \dfrac{4 \ \cancel{\text{qt}}}{1 \ \cancel{\text{gal}}} \times \dfrac{1 \text{ L}}{1.06 \ \cancel{\text{qt}}}$

$= \dfrac{14 \times 4}{1.06} \text{ L}$

$\approx 52.83 \text{ L}$

29. 10 is 50% of what number?

$10 = 0.50 \cdot n$

$\dfrac{10}{0.50} = \dfrac{\cancel{0.50} \cdot n}{\cancel{0.50}}$

$20 = n$

31. Difference = High − Low

$$= 9° - (-8°)$$
$$= 9° + 8°$$
$$= 17°$$

33. Let x = number of women

$$\frac{3 \text{ men}}{4 \text{ women}} = \frac{15 \text{ men}}{x \text{ women}}$$
$$3 \cdot x = 4 \cdot 15$$
$$3x = 60$$
$$x = 20$$

There are 20 women in the class.

35. $\text{Average (mean)} = \dfrac{\text{sum of the values}}{\text{number of values}}$

$$= \frac{72 + 113 + 108 + 95}{4}$$
$$= \frac{388}{4}$$
$$= 97 \text{ miles}$$

37.

Cost per pound	×	Number of pounds	=	Total Cost
$4.32	×	2.5	=	$10.80

39. Let x = number of two point baskets

$$\frac{8 \text{ baskets}}{18 \text{ games}} = \frac{x \text{ baskets}}{45 \text{ games}}$$
$$8 \cdot 45 = 18 \cdot x$$
$$360 = 18x$$
$$20 = x$$

If she continues at the same rate, she will make 20 two point baskets.

Chapter 7 Test

1. The opposite of 14 is –14.

3. $-1 > -4$, because –1 is right of –4 on the number line.

5. $-(-7) = 7$
The opposite of –7 is 7.

7. $8 + (-17) = -9$

9. $-\dfrac{2}{3} + \left(-\dfrac{4}{5}\right) = -\dfrac{2 \cdot 5}{3 \cdot 5} + \left(-\dfrac{4 \cdot 3}{5 \cdot 3}\right)$

$= -\dfrac{10}{15} + \left(-\dfrac{12}{15}\right)$

$= -\dfrac{22}{15}$ or $-1\dfrac{7}{15}$

11. $(-6)(-7) = 42$

13. $\dfrac{-80}{16} = -5$

15. $(-3)^2 = (-3)(-3)$
$= 9$

17. $(-7)(3) + (-2)(-5)$ *multiply before add*
$= -21 + 10$ *add*
$= -11$

19. Simplify above and below fraction bar first:
$\dfrac{-5 + 3(-3)}{5 - 7}$

$= \dfrac{-5 + (-9)}{5 - 7}$ *add and subtract*

$= \dfrac{-14}{-2}$ *divide*

$= 7$

21. $-15 + (-46) = -61$

23. $(-8)(-3) = 24$

25. Above each year, draw a dot corresponding to the amount of garbage for that year as measured on the vertical axis. To draw the line graph, connect the dots.

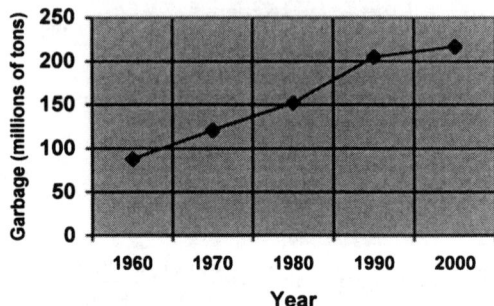

27. To find the difference, subtract the low temperature from the high temperature:
$21° - (-4°) = 21° + 4°$
$\qquad\qquad = 25°$

29. $4(5x - 1) = 4 \cdot 5x - 4 \cdot 1$
$= (4 \cdot 5)x - 4$
$= 20x - 4$

31. $12x + 20x = (12 + 20)x = 32x$

Chapter 8 Preview

1. $-2 + 7 = 5$

3. $-2 + (-4) = -6$

5. $(-4)(5) = -20$

7. $\dfrac{3}{2}(12) = \dfrac{3}{2} \cdot \dfrac{12}{1} = \dfrac{3 \cdot \cancel{2} \cdot 6}{\cancel{2} \cdot 1} = 18$

9. $\left(-\dfrac{5}{4}\right)\left(\dfrac{8}{15}\right) = -\dfrac{\cancel{5} \cdot 2 \cdot \cancel{4}}{\cancel{4} \cdot 3 \cdot \cancel{5}} = -\dfrac{2}{3}$

11. $\dfrac{1}{3}(15) + 2 = 5 + 2 = 7$

13. $3x + 7x = (3 + 7)x = 10x$

15. $4(x - 5) = 4 \cdot x - 4 \cdot 5 = 4x - 20$

17. $6x - x = (6 - 1)x = 5x$

19. $P = 2l + 2w$
$= 2(3x) + 2(x)$
$= 6x + 2x$
$= 8x$

Chapter 8 Pretest

1. $4a - 1 - 5a + 8 = 4a - 5a - 1 + 8$

$$= (4 - 5)a - 1 + 8$$

$$= -a + 7$$

3. When $x = -2$:

$$3x - 4 = 3(-2) - 4$$

$$= -6 - 4$$

$$= -10$$

5. $\qquad a + 4 = -2$

$$a + 4 + (-4) = -2 + (-4)$$

$$a = -6$$

7. $\qquad \dfrac{1}{5}x = 3$

$$5\left(\dfrac{1}{5}x\right) = 5(3)$$

$$x = 15$$

9. $\qquad 2a + 1 = 5(a - 2) - 1$

$$2a + 1 = 5a - 10 - 1$$

$$2a + 1 = 5a - 11$$

$$2a + 1 + (-2a) = 5a - 11 + (-2a)$$

$$1 = 3a - 11$$

$$1 + 11 = 3a - 11 + 11$$

$$12 = 3a$$

$$4 = a$$

11. Let x = number asked for

$\underbrace{\text{The sum of a number and 6}}$ is -17

$\qquad\qquad \downarrow \qquad\qquad \downarrow \quad \downarrow$

$\qquad\qquad x + 6 \qquad\quad = \ -17$

$x + 6 = -17$

$\qquad x = -23$

The number is -23.

13. Let w = width.

Then $w + 4$ = length.

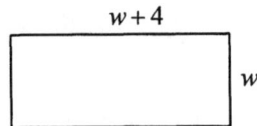

$w + 4$

w

$P = 2l + 2w$

$44 = 2(w + 4) + 2w$

$44 = 2w + 8 + 2w$

$44 = 4w + 8$

$36 = 4w$

$9 = w$

The width is $w = 9$ in.

The length is $w + 4 = 9 + 4 = 13$ in.

Section 8.1

1. $2x + 8x = (2+8)x = 10x$

3. $6a - 2a = (6-2)a = 4a$

5. $-4y + 5y = (-4+5)y = 1y = y$

7. $-6x - 2x = (-6-2)x = -8x$

9. $4a - a = 4a - 1a = (4-1)a = 3a$

11. $x - 6x = 1x - 6x = (1-6)x = -5x$

13. $4x + 2x + 3 + 8 = (4+2)x + 3 + 8$
$$= 6x + 11$$

15. $7x - 5x + 6 - 4 = (7-5)x + 6 - 4$
$$= 2x + 2$$

17. $-2a + a + 7 + 5 = (-2+1)a + 7 + 5$
$$= -1a + 12$$
$$= -a + 12$$

19. $6y - 2y - 5 + 1 = (6-2)y - 5 + 1$
$$= 4y - 4$$

21. $4x + 2x - 8x + 4 = (4+2-8)x + 4$
$$= -2x + 4$$

23. $9x - x - 5 - 1 = (9-1)x - 5 - 1$
$$= 8x - 6$$

25. $2a + 4 + 3a + 5 = 2a + 3a + 4 + 5$
$$= (2+3)a + 9$$
$$= 5a + 9$$

27. $3x + 2 - 4x + 1 = 3x - 4x + 2 + 1$
$$= (3-4)x + 2 + 1$$
$$= -1x + 3$$
$$= -x + 3$$

29. $12y + 3 + 5y = 12y + 5y + 3$
$$= (12+5)y + 3$$
$$= 17y + 3$$

31. $4a - 3 - 5a + 2a = 4a - 5a + 2a - 3$
$$= (4-5+2)a - 3$$
$$= 1a - 3$$
$$= a - 3$$

33. $2(3x+4) + 8 = 2(3x) + 2(4) + 8$
$$= 6x + 8 + 8$$
$$= 6x + 16$$

35. $5(2x-3) + 4 = 5(2x) - 5(3) + 4$
$$= 10x - 15 + 4$$
$$= 10x - 11$$

37. $8(2y+4) + 3y = 8(2y) + 8(4) + 3y$
$$= 16y + 32 + 3y$$
$$= 16y + 3y + 32$$
$$= 19y + 32$$

39. $6(4y-3) + 6y = 6(4y) - 6(3) + 6y$
$$= 24y - 18 + 6y$$
$$= 24y + 6y - 18$$
$$= 30y - 18$$

41. $2(x+3) + 4(x+2)$
$$= 2(x) + 2(3) + 4(x) + 4(2)$$
$$= 2x + 6 + 4x + 8$$
$$= 2x + 4x + 6 + 8$$
$$= 6x + 14$$

43. $3(2a+4)+7(3a-1)$
$$= 3(2a)+3(4)+7(3a)+7(-1)$$
$$= 6a+12+21a-7$$
$$= 6a+21a+12-7$$
$$= 27a+5$$

45. When $x = 5$:
$$2x+4 = 2(5)+4$$
$$= 10+4$$
$$= 14$$

47. When $x = 5$:
$$7x-8 = 7(5)-8$$
$$= 35-8$$
$$= 27$$

49. When $x = 5$:
$$-4x+1 = -4(5)+1$$
$$= -20+1$$
$$= -19$$

51. When $x = 5$:
$$-8+3x = -8+3(5)$$
$$= -8+15$$
$$= 7$$

53. When $a = -2$:
$$2a+5 = 2(-2)+5$$
$$= -4+5$$
$$= 1$$

55. When $a = -2$:
$$-7a+4 = -7(-2)+4$$
$$= 14+4$$
$$= 18$$

57. When $a = -2$:
$$-a+10 = -(-2)+10$$
$$= 2+10$$
$$= 12$$

59. When $a = -2$:
$$-4+3a = -4+3(-2)$$
$$= -4+(-6)$$
$$= -10$$

61. When $x = 3$:
$$3x+5x+4 = 3(3)+5(3)+4$$
$$= 9+15+4$$
$$= 28$$

63. When $x = 3$:
$$9x+x+3+7 = 9(3)+(3)+3+7$$
$$= 27+3+3+7$$
$$= 40$$

65. When $x = 3$:
$$4x+3+2x+5 = 4(3)+3+2(3)+5$$
$$= 12+3+6+5$$
$$= 26$$

67. When $x = 3$:
$$3x-8+2x-3 = 3(3)-8+2(3)-3$$
$$= 9-8+6-3$$
$$= 4$$

69. $\text{Area} = 6(x)+6(4) \quad \text{Area} = 6(x+4)$
$$= 6x+24 \qquad\qquad = 6x+24$$

71. $P = 4s = 4(x+1) = 4x+4$

73. $P = 2a+2b$
$$= 2(3x+1)+2(2x-3)$$
$$= 6x+2+4x-6$$
$$= 10x-4$$

Applying the Concepts

75. Complement = $90° - 25° = 65°$
Supplement = $180° - 25° = 155°$
$25°$ is an acute angle, because it is between $0°$ and $90°$.

77. a. When $A = 12,000$ ft:

$$72 - \frac{A}{300} = 72 - \frac{12,000}{300}$$
$$= 72 - 40$$
$$= 32$$

The temperature is 32°F.

b. When $A = 15,000$ ft:

$$72 - \frac{A}{300} = 72 - \frac{15,000}{300}$$
$$= 72 - 50$$
$$= 22$$

The temperature is 22°F.

c. When $A = 27,000$ ft:

$$72 - \frac{A}{300} = 72 - \frac{27,000}{300}$$
$$= 72 - 90$$
$$= -18$$

The temperature is −18°F.

79. a. When $g = 10$ gallons:

$$7 + 2g = 7 + 2(10)$$
$$= 7 + 20$$
$$= 27$$

The monthly bill is $27.

b. When $g = 20$ gallons:

$$7 + 2g = 7 + 2(20)$$
$$= 7 + 40$$
$$= 47$$

The monthly bill is $47.

Review Problems

81. The opposite of 9 is −9.

83. The opposite of −6 is $-(-6) = 6$.

85. $2(6+5) = 2(6) + 2(5)$
Distributive Property

87. $x + 5 = 5 + x$
Commutative Property of Addition

89. $(x+5) + 1 = 1 + (x+5)$
Commutative Property of Addition

Section 8.2

1. When $x = 2$:
$$2x + 1 = 5$$
$$2(2) + 1 = 5$$
$$4 + 1 = 5$$
$$5 = 5 \quad \text{true}$$
$x = 2$ is a solution.

3. When $x = 5$:
$$3x + 4 = 19$$
$$3(5) + 4 = 19$$
$$15 + 4 = 19$$
$$19 = 19 \quad \text{true}$$
$x = 5$ is a solution.

5. When $x = 4$:
$$2x - 4 = 2$$
$$2(4) - 4 = 2$$
$$8 - 4 = 2$$
$$4 = 2 \quad \text{false}$$
$x = 4$ is not a solution.

7. When $x = -2$:
$$2x + 1 = 3x + 3$$
$$2(-2) + 1 = 3(-2) + 3$$
$$-4 + 1 = -6 + 3$$
$$-3 = -3 \quad \text{true}$$
$x = -2$ is a solution.

9. When $x = -4$:
$$x - 4 = 2x + 1$$
$$(-4) - 4 = 2(-4) + 1$$
$$-8 = -8 + 1$$
$$-8 = -7 \quad \text{false}$$
$x = -4$ is not a solution.

11.
$$x + 2 = 8$$
$$x + 2 + (-2) = 8 + (-2)$$
$$x + 0 = 6$$
$$x = 6$$

13.
$$x - 4 = 7$$
$$x - 4 + 4 = 7 + 4$$
$$x - 0 = 11$$
$$x = 11$$

15.
$$a + 9 = -6$$
$$a + 9 + (-9) = -6 + (-9)$$
$$a + 0 = -15$$
$$a = -15$$

17.
$$x - 5 = -4$$
$$x - 5 + 5 = -4 + 5$$
$$x - 0 = 1$$
$$x = 1$$

19.
$$y - 3 = -6$$
$$y - 3 + 3 = -6 + 3$$
$$y - 0 = -3$$
$$y = -3$$

21.
$$a + \frac{1}{3} = -\frac{2}{3}$$
$$a + \frac{1}{3} + \left(-\frac{1}{3}\right) = -\frac{2}{3} + \left(-\frac{1}{3}\right)$$
$$a + 0 = -\frac{3}{3}$$
$$a = -1$$

23.
$$x - \frac{3}{5} = \frac{4}{5}$$
$$x - \frac{3}{5} + \frac{3}{5} = \frac{4}{5} + \frac{3}{5}$$
$$x - 0 = \frac{7}{5}$$
$$x = \frac{7}{5}$$

25.
$$y + 7.3 = -2.7$$
$$y + 7.3 + (-7.3) = -2.7 + (-7.3)$$
$$y + 0 = -10$$
$$y = -10$$

27. $x + 4 - 7 = 3 - 10$
$$x - 3 = -7$$
$$x - 3 + 3 = -7 + 3$$
$$x = -4$$

29. $x - 6 + 4 = -3 - 2$
$$x - 2 = -5$$
$$x - 2 + 2 = -5 + 2$$
$$x = -3$$

31. $3 - 5 = a - 4$
$$-2 = a - 4$$
$$-2 + 4 = a - 4 + 4$$
$$2 = a$$

33. $3a + 7 - 2a = 1$
$$a + 7 = 1$$
$$a + 7 + (-7) = 1 + (-7)$$
$$a = -6$$

35. $6a - 2 - 5a = -9 + 1$
$$a - 2 = -8$$
$$a - 2 + 2 = -8 + 2$$
$$a = -6$$

37. $8 - 5 = 3x - 2x + 4$
$$3 = x + 4$$
$$3 + (-4) = x + 4 + (-4)$$
$$-1 = x$$

39. $2(x + 3) - x = 4$
$$2x + 6 - x = 4$$
$$x + 6 = 4$$
$$x + 6 + (-6) = 4 + (-6)$$
$$x = -2$$

41. $-3(x - 4) + 4x = 3 - 7$
$$-3x + 12 + 4x = -4$$
$$x + 12 = -4$$
$$x + 12 + (-12) = -4 + (-12)$$
$$x = -16$$

43. $5(2a + 1) - 9a = 8 - 6$
$$10a + 5 - 9a = 2$$
$$a + 5 = 2$$
$$a + 5 + (-5) = 2 + (-5)$$
$$a = -3$$

45. $-(x + 3) + 2x - 1 = 6$
$$-x - 3 + 2x - 1 = 6$$
$$x - 4 = 6$$
$$x - 4 + 4 = 6 + 4$$
$$x = 10$$

47. $P = $ sum of sides
$$36 = 10 + 10 + x + 12$$
$$36 = x + 32$$
$$36 + (-32) = x + 32 + (-32)$$
$$4 = x$$

49. $P = $ sum of sides
$$16 = 5 + 5 + x - 6$$
$$16 = x + 4$$
$$12 = x$$

Applying the Concepts

51.
$$x + 23° = 90°$$
$$x + 23° + (-23°) = 90° + (-23°)$$
$$x = 67°$$

53. a.
$$x + 86 + 89 = 400$$
$$x + 175 = 400$$
$$x + 175 + (-175) = 400 + (-175)$$
$$x = 225$$
There are 225 on the stage level.

b. $30 \times$ stage seats $+ \$25 \times$ other seats $=$ maximum amount of money
$$30(225) + 25(86 + 89)$$
$$= 30(225) + 25(175)$$
$$= 6,750 + 4,375$$
$$= 11,125$$
The maximum amount of money is $11,125.

55.
$$x + 12 = 30$$
$$x + 12 + (-12) = 30 + (-12)$$
$$x = 18$$

57.

The difference of	is equal to	the sum of
8 and 5		x and 7

$$8 - 5 \quad = \quad x + 7$$
$$8 - 5 = x + 7$$
$$3 = x + 7$$
$$3 + (-7) = x + 7 + (-7)$$
$$-4 = x$$

Review Problems

59.
$$\frac{x}{2} = \frac{5}{10}$$
$$10 \cdot x = 2 \cdot 5$$
$$10x = 10$$
$$\frac{10x}{10} = \frac{10}{10}$$
$$x = 1$$

61.
$$\frac{x}{7} = \frac{1}{14}$$
$$14 \cdot x = 7 \cdot 1$$
$$14x = 7$$
$$\frac{14x}{14} = \frac{7}{14}$$
$$x = \frac{7}{14} = \frac{1}{2}$$

63.
$$\frac{x}{3} = \frac{2}{5}$$
$$5 \cdot x = 3 \cdot 2$$
$$5x = 6$$
$$\frac{5x}{5} = \frac{6}{5}$$
$$x = \frac{6}{5} = 1\frac{1}{5}$$

65.
$$\frac{x}{6} = \frac{4}{3}$$
$$3 \cdot x = 6 \cdot 4$$
$$3x = 24$$
$$\frac{3x}{3} = \frac{24}{3}$$
$$x = 8$$

67.

$$\frac{x}{6} = \frac{6}{5}$$

$$5 \cdot x = 6 \cdot 6$$

$$5x = 36$$

$$\frac{5x}{5} = \frac{36}{5}$$

$$x = 7\frac{1}{5}$$

Section 8.3

1. $\dfrac{1}{4}x = 2$

$4 \cdot \dfrac{1}{4}x = 4 \cdot 2$

$x = 8$

3. $\dfrac{1}{2}x = -3$

$2\left(\dfrac{1}{2}x\right) = 2(-3)$

$x = -6$

5. $-\dfrac{1}{3}x = 2$

$-3\left(-\dfrac{1}{3}x\right) = -3(2)$

$x = -6$

7. $-\dfrac{1}{6}x = -1$

$-6\left(-\dfrac{1}{6}x\right) = -6(-1)$

$x = 6$

9. $\dfrac{3}{4}y = 12$

$\dfrac{4}{3} \cdot \dfrac{3}{4}y = \dfrac{4}{3} \cdot 12$

$y = 16$

11. $3a = 48$

$\dfrac{3a}{3} = \dfrac{48}{3}$

$a = 16$

13. $-\dfrac{3}{5}x = \dfrac{9}{10}$

$-\dfrac{5}{3}\left(-\dfrac{3}{5}x\right) = -\dfrac{5}{3}\left(\dfrac{9}{10}\right)$

$x = -\dfrac{3}{2}$

15. $5x = -35$

$\dfrac{5x}{5} = \dfrac{-35}{5}$

$x = -7$

17. $-8y = 64$

$\dfrac{-8y}{-8} = \dfrac{64}{-8}$

$y = -8$

19. $-7x = -42$

$\dfrac{-7x}{-7} = \dfrac{-42}{-7}$

$x = 6$

21. $3x - 1 = 5$

$3x - 1 + 1 = 5 + 1$

$3x = 6$

$\dfrac{3x}{3} = \dfrac{6}{3}$

$x = 2$

23. $-4a + 3 = -9$

$-4a + 3 + (-3) = -9 + (-3)$

$-4a = -12$

$\dfrac{-4a}{-4} = \dfrac{-12}{-4}$

$a = 3$

25.
$$6x - 5 = 19$$
$$6x - 5 + 5 = 19 + 5$$
$$6x = 24$$
$$\frac{6x}{6} = \frac{24}{6}$$
$$x = 4$$

27.
$$\frac{1}{3}a + 3 = -5$$
$$\frac{1}{3}a + 3 + (-3) = -5 + (-3)$$
$$\frac{1}{3}a = -8$$
$$3\left(\frac{1}{3}a\right) = 3(-8)$$
$$a = -24$$

29.
$$-\frac{1}{4}a + 5 = 2$$
$$-\frac{1}{4}a + 5 + (-5) = 2 + (-5)$$
$$-\frac{1}{4}a = -3$$
$$-4\left(-\frac{1}{4}a\right) = -4(-3)$$
$$a = 12$$

31.
$$2x - 4 = -20$$
$$2x - 4 + 4 = -20 + 4$$
$$2x = -16$$
$$\frac{2x}{2} = \frac{-16}{2}$$
$$x = -8$$

33.
$$\frac{2}{3}x - 4 = 6$$
$$\frac{2}{3}x - 4 + 4 = 6 + 4$$
$$\frac{2}{3}x = 10$$
$$\frac{3}{2} \cdot \frac{2}{3}x = \frac{3}{2} \cdot 10$$
$$x = 15$$

35.
$$-11a + 4 = -29$$
$$-11a + 4 + (-4) = -29 + (-4)$$
$$-11a = -33$$
$$\frac{-11a}{-11} = \frac{-33}{-11}$$
$$a = 3$$

37.
$$-3y - 2 = 1$$
$$-3y - 2 + 2 = 1 + 2$$
$$-3y = 3$$
$$\frac{-3y}{-3} = \frac{3}{-3}$$
$$y = -1$$

39.
$$-2x - 5 = -7$$
$$-2x - 5 + 5 = -7 + 5$$
$$-2x = -2$$
$$\frac{-2x}{-2} = \frac{-2}{-2}$$
$$x = 1$$

41.
$$2x + 3x - 5 = 7 + 3$$
$$5x - 5 = 10$$
$$5x - 5 + 5 = 10 + 5$$
$$5x = 15$$
$$\frac{5x}{5} = \frac{15}{5}$$
$$x = 3$$

43. $4x - 7 + 2x = 9 - 10$

$$6x - 7 = -1$$
$$6x - 7 + 7 = -1 + 7$$
$$6x = 6$$
$$\frac{6x}{6} = \frac{6}{6}$$
$$x = 1$$

45. $3a + 2a + a = 7 - 13$

$$6a = -6$$
$$\frac{6a}{6} = \frac{-6}{6}$$
$$a = -1$$

47. $5x + 4x + 3x = 4 - 8$

$$12x = -4$$
$$\frac{12x}{12} = \frac{-4}{12}$$
$$x = -\frac{1}{3}$$

49.

$$5 - 18 = 3y - 2y + 1$$
$$-13 = y + 1$$
$$-13 + (-1) = y + 1 + (-1)$$
$$-14 = y$$

51. $P = 4s$

$$72 = 4(2x)$$
$$72 = 8x$$
$$\frac{72}{8} = \frac{8x}{8}$$
$$9 = x$$

53. $P = 2l + 2w$

$$80 = 2(3x) + 2(2x)$$
$$80 = 6x + 4x$$
$$80 = 10x$$
$$\frac{80}{10} = \frac{10x}{10}$$
$$8 = x$$

Applying the Concepts

55. $1 + 2(4) + 3x = 21$

$$9 + 3x = 21$$
$$9 + (-9) + 3x = 21 + (-9)$$
$$3x = 12$$
$$\frac{3x}{3} = \frac{12}{3}$$
$$x = 4$$

She made 4 three point shots.

57. $32x - 25x - 18 = 2,082$

$$7x - 18 = 2,082$$
$$7x - 18 + 18 = 2,082 + 18$$
$$7x = 2,100$$
$$\frac{7x}{7} = \frac{2,100}{7}$$
$$x = 300$$

You originally purchased 300 shares.

59. The sum of $2x$ and 5 is 19.

$$2x + 5 \qquad = 19$$
$$2x + 5 = 19$$
$$2x + 5 + (-5) = 19 + (-5)$$
$$2x = 14$$
$$\frac{2x}{2} = \frac{14}{2}$$
$$x = 7$$

61. The difference of $5x$ and 6 is -9.

$$5x - 6 \qquad = -9$$

$$5x - 6 = -9$$

$$5x - 6 + 6 = -9 + 6$$

$$5x = -3$$

$$\frac{5x}{5} = \frac{-3}{5}$$

$$x = -\frac{3}{5}$$

Review Problems

63. $2(3a - 8) = 2 \cdot 3a - 2 \cdot 8$

$$= 6a - 16$$

65. $-3(5x - 1) = -3(5x) + (-3)(-1)$

$$= -15x + 3$$

67. $3(y - 5) + 6 = 3 \cdot y - 3 \cdot 5 + 6$

$$= 3y - 15 + 6$$

$$= 3y - 9$$

69. $6(2x - 1) + 4x = 6 \cdot 2x - 6 \cdot 1 + 4x$

$$= 12x - 6 + 4x$$

$$= 12x + 4x - 6$$

$$= 16x - 6$$

Section 8.4

1.
$$5(x+1) = 20$$
$$5x + 5 = 20$$
$$5x + 5 + (-5) = 20 + (-5)$$
$$5x = 15$$
$$\frac{5x}{5} = \frac{15}{5}$$
$$x = 3$$

3.
$$6(x-3) = -6$$
$$6x - 18 = -6$$
$$6x - 18 + 18 = -6 + 18$$
$$6x = 12$$
$$\frac{6x}{6} = \frac{12}{6}$$
$$x = 2$$

5.
$$2x + 4 = 3x + 7$$
$$2x + 4 + (-3x) = 3x + 7 + (-3x)$$
$$-1x + 4 = 7$$
$$-1x + 4 + (-4) = 7 + (-4)$$
$$-1x = 3$$
$$\frac{-1x}{-1} = \frac{3}{-1}$$
$$x = -3$$

7.
$$7y - 3 = 4y - 15$$
$$7y - 3 + (-4y) = 4y - 15 + (-4y)$$
$$3y - 3 = -15$$
$$3y - 3 + 3 = -15 + 3$$
$$3y = -12$$
$$\frac{3y}{3} = \frac{-12}{3}$$
$$y = -4$$

9.
$$12x + 3 = -2x + 17$$
$$12x + 3 + 2x = -2x + 17 + 2x$$
$$14x + 3 = 17$$
$$14x + 3 + (-3) = 17 + (-3)$$
$$14x = 14$$
$$\frac{14x}{14} = \frac{14}{14}$$
$$x = 1$$

11.
$$6x - 8 = -x - 8$$
$$6x - 8 + x = -x - 8 + x$$
$$7x - 8 = -8$$
$$7x - 8 + 8 = -8 + 8$$
$$7x = 0$$
$$\frac{7x}{7} = \frac{0}{7}$$
$$x = 0$$

13.
$$7(a-1) + 4 = 11$$
$$7a - 7 + 4 = 11$$
$$7a - 3 = 11$$
$$7a - 3 + 3 = 11 + 3$$
$$7a = 14$$
$$\frac{7a}{7} = \frac{14}{7}$$
$$a = 2$$

15.
$$8(x+5) - 6 = 18$$
$$8x + 40 - 6 = 18$$
$$8x + 34 = 18$$
$$8x + 34 + (-34) = 18 + (-34)$$
$$8x = -16$$
$$\frac{8x}{8} = \frac{-16}{8}$$
$$x = -2$$

17. $2(3x-6)+1=7$

$\qquad 6x-12+1=7$

$\qquad 6x-11=7$

$\qquad 6x-11+11=7+11$

$\qquad 6x=18$

$\qquad \dfrac{6x}{6}=\dfrac{18}{6}$

$\qquad x=3$

19. $10(y+1)+4=3y+7$

$\qquad 10y+10+4=3y+7$

$\qquad 10y+14=3y+7$

$\qquad 10y+14+(-3y)=3y+7+(-3y)$

$\qquad 7y+14=7$

$\qquad 7y+14+(-14)=7+(-14)$

$\qquad 7y=-7$

$\qquad \dfrac{7y}{7}=\dfrac{-7}{7}$

$\qquad y=-1$

21. $4(x-6)+1=2x-9$

$\qquad 4x-24+1=2x-9$

$\qquad 4x-23=2x-9$

$\qquad 4x-23+(-2x)=2x-9+(-2x)$

$\qquad 2x-23=-9$

$\qquad 2x-23+23=-9+23$

$\qquad 2x=14$

$\qquad \dfrac{2x}{2}=\dfrac{14}{2}$

$\qquad x=7$

23. $2(3x+1)=4(x-1)$

$\qquad 6x+2=4x-4$

$\qquad 6x+2+(-4x)=4x-4+(-4x)$

$\qquad 2x+2=-4$

$\qquad 2x+2+(-2)=-4+(-2)$

$\qquad 2x=-6$

$\qquad \dfrac{2x}{2}=\dfrac{-6}{2}$

$\qquad x=-3$

25. $3a+4=2(a-5)+15$

$\qquad 3a+4=2a-10+15$

$\qquad 3a+4=2a+5$

$\qquad 3a+4+(-2a)=2a+5+(-2a)$

$\qquad a+4=5$

$\qquad a+4+(-4)=5+(-4)$

$\qquad a=1$

27. $9x-6=-3(x+2)-24$

$\qquad 9x-6=-3x-6-24$

$\qquad 9x-6=-3x-30$

$\qquad 9x-6+3x=-3x-30+3x$

$\qquad 12x-6=-30$

$\qquad 12x-6+6=-30+6$

$\qquad 12x=-24$

$\qquad \dfrac{12x}{12}=\dfrac{-24}{12}$

$\qquad x=-2$

29.
$$3x - 5 = 11 + 2(x - 6)$$
$$3x - 5 = 11 + 2x - 12$$
$$3x - 5 = -1 + 2x$$
$$3x - 5 + (-2x) = -1 + 2x + (-2x)$$
$$x - 5 = -1$$
$$x - 5 + 5 = -1 + 5$$
$$x = 4$$

31.
$$\frac{x}{3} + \frac{x}{6} = 5 \qquad \text{LCD} = 6$$
$$6\left(\frac{x}{3} + \frac{x}{6}\right) = 6(5)$$
$$6\left(\frac{x}{3}\right) + 6\left(\frac{x}{6}\right) = 6(5)$$
$$2x + x = 30$$
$$3x = 30$$
$$x = 10$$

33.
$$\frac{x}{5} - x = 4 \qquad \text{LCD} = 5$$
$$5\left(\frac{x}{5} - x\right) = 5(4)$$
$$5\left(\frac{x}{5}\right) - 5(x) = 5(4)$$
$$x - 5x = 20$$
$$-4x = 20$$
$$x = -5$$

35.
$$3x + \frac{1}{2} = \frac{1}{4} \qquad \text{LCD} = 4$$
$$4\left(3x + \frac{1}{2}\right) = 4\left(\frac{1}{4}\right)$$
$$4(3x) + 4\left(\frac{1}{2}\right) = 4\left(\frac{1}{4}\right)$$
$$12x + 2 = 1$$
$$12x = -1$$
$$x = -\frac{1}{12}$$

37.
$$\frac{x}{3} + \frac{1}{2} = -\frac{1}{2} \qquad \text{LCD} = 6$$
$$6\left(\frac{x}{3} + \frac{1}{2}\right) = 6\left(-\frac{1}{2}\right)$$
$$6\left(\frac{x}{3}\right) + 6\left(\frac{1}{2}\right) = 6\left(-\frac{1}{2}\right)$$
$$2x + 3 = -3$$
$$2x = -6$$
$$x = -3$$

39.
$$\frac{4}{x} = \frac{1}{5} \qquad \text{LCD} = 5x$$
$$5x\left(\frac{4}{x}\right) = 5x\left(\frac{1}{5}\right)$$
$$20 = x$$

41.
$$\frac{3}{x} + 1 = \frac{2}{x} \qquad \text{LCD} = x$$
$$x\left(\frac{3}{x} + 1\right) = x\left(\frac{2}{x}\right)$$
$$x\left(\frac{3}{x}\right) + x(1) = x\left(\frac{2}{x}\right)$$
$$3 + x = 2$$
$$x = -1$$

43.
$$\frac{3}{x} - \frac{2}{x} = \frac{1}{5} \qquad LCD = 5x$$

$$5x\left(\frac{3}{x} - \frac{2}{x}\right) = 5x\left(\frac{1}{5}\right)$$

$$5x\left(\frac{3}{x}\right) - 5x\left(\frac{2}{x}\right) = 5x\left(\frac{1}{5}\right)$$

$$15 - 10 = x$$

$$5 = x$$

45.
$$\frac{1}{x} - \frac{1}{2} = -\frac{1}{4} \qquad LCD = 4x$$

$$4x\left(\frac{1}{x} - \frac{1}{2}\right) = 4x\left(-\frac{1}{4}\right)$$

$$4x\left(\frac{1}{x}\right) - 4x\left(\frac{1}{2}\right) = 4x\left(-\frac{1}{4}\right)$$

$$4 - 2x = -x$$

$$4 - 2x + 2x = -x + 2x$$

$$4 = x$$

47.
$$4x - 4.7 = 3.5$$

$$4x - 4.7 + 4.7 = 3.5 + 4.7$$

$$4x = 8.2$$

$$\frac{4x}{4} = \frac{8.2}{4}$$

$$x = 2.05$$

49.
$$0.02 + 5y = -0.3$$

$$0.02 + 5y + (-0.02) = -0.3 + (-0.02)$$

$$5y = -0.32$$

$$\frac{5y}{5} = -\frac{0.32}{5}$$

$$y = -0.064$$

51.
$$\frac{1}{3}x - 2.99 = 1.02$$

$$\frac{1}{3}x - 2.99 + 2.99 = 1.02 + 2.99$$

$$\frac{1}{3}x = 4.01$$

$$3\left(\frac{1}{3}x\right) = 3(4.01)$$

$$x = 12.03$$

53.
$$7n - 0.32 = 5n + 0.56$$

$$7n - 0.32 + 0.32 = 5n + 0.56 + 0.32$$

$$7n = 5n + 0.88$$

$$7n + (-5n) = 5n + 0.88 + (-5n)$$

$$2n = 0.88$$

$$n = 0.44$$

55.
$$3a + 4.6 = 7a + 5.3$$

$$3a + 4.6 + (-4.6) = 7a + 5.3 + (-4.6)$$

$$3a = 7a + 0.7$$

$$3a + (-7a) = 7a + 0.7 + (-0.7)$$

$$-4a = 0.7$$

$$a = -0.175$$

57.
$$0.5x + 0.1(x + 20) = 3.2$$

$$0.5x + 0.1x + 2 = 3.2$$

$$0.6x + 2 = 3.2$$

$$0.6x + 2 + (-2) = 3.2 + (-2)$$

$$0.6x = 1.2$$

$$x = 2$$

59. P = sum of sides:
$$x + x + x + 6 = 36$$

$$3x + 6 = 36$$

$$3x = 30$$

$$x = 10$$

61. $P = $ sum of sides:

$$x + x + x + 1 = 16$$
$$3x + 1 = 16$$
$$3x = 15$$
$$x = 5$$

Applying the Concepts

63.
$$58 - x = 45$$
$$58 - x + (-58) = 45 + (-58)$$
$$-x = -13$$
$$\frac{-1x}{-1} = \frac{-13}{-1}$$
$$x = 13$$

$$45 + x = 58$$
$$45 + x + (-45) = 58 + (-45)$$
$$x = 13$$

Solving either equation, the angle is 13°.

65.
$$x + \frac{1}{4}x = 15 \qquad \text{LCD} = 4$$
$$4\left(x + \frac{1}{4}x\right) = 4(15)$$
$$4(x) + 4\left(\frac{1}{4}x\right) = 4(15)$$
$$4x + x = 60$$
$$5x = 60$$
$$x = 12$$

The quantity is 12.

Review Problems

67. The sum of

a number and 2

$$x \qquad + \quad 2$$

69. Twice a number

$$2 \cdot \qquad x \qquad = 2x$$

71. Twice   the sum of

a number and 6

$$2 \cdot \qquad (x + 6) \qquad = 2(x + 6)$$

73. The difference of

x and 4

$$x - 4$$

75. The sum of

twice a number and 5

$$2x \qquad + \quad 5$$

Extending the Concepts

77. a. Let $x = $ number of general tickets.
   Then $x + 100 = $ number of student tickets.

   Student sales + general sales = \$2,400

   $$\$4(x + 100) \quad + \quad \$6x \quad = \$2,400$$

   b. $4(x + 100) + 6x = 2,400$
   $$4x + 400 + 6x = 2,400$$
   $$10x + 400 = 2,400$$
   $$10x = 2,000$$
   $$x = 200$$

   There were $x = 200$ general tickets, and
   $x + 100 = 200 + 100 = 300$ student tickets.

   c. Total tickets: $300 + 200 = 500$.

Section 8.5

1. The sum of
$$\underbrace{x \text{ and } 3}$$

$$x + 3$$

3. The sum of
$$\underbrace{\text{twice } x} \text{ and } 1$$

$$2x \quad + \quad 1$$

5. $\underbrace{\text{Five } x} \,\, \underbrace{\text{decreased by}} \,\, 6$

$$5x \quad\quad - \quad\quad 6$$

7. $\underbrace{\text{Three times}} \quad \underbrace{\text{the sum of} \atop x \text{ and } 1}$

$$3\times \quad\quad (x+1) \quad = 3(x+1)$$

9. $\underbrace{\text{Five times}} \quad \underbrace{{\text{the sum of} \atop \text{three } x \text{ and } 4}}$

$$5\times \quad\quad (3x+4) \quad = 5(3x+4)$$

Applying the Concepts

11. Step 1: *Read and list*
  Known items: the numbers 3 and 5.
  Unknown items: the number in question
  Step 2: *Assign a variable and translate*
  Let x = the number asked for
  Then $x + 3$ ="the sum of a number and 3".
  Step 3: *Reread and write an equation*
  $\underbrace{\text{The sum of } x \text{ and } 3}$ is 5.

$$\downarrow \quad\quad\quad \downarrow\downarrow$$
$$x + 3 \quad\quad = 5$$

  Step 4: *Solve the equation*
  $x + 3 = 5$
  $x = 2$
  Step 5: *Write your answer*
  The number is 2.
  Step 6: *Reread and check*
  The sum of 2 and 3 is 5.

13. Step 1: *Read and list*
  Known items: the numbers 1 and −3, twice a number.
  Unknown items: the number in question
  Step 2: *Assign a variable and translate*
  Let x = the number asked for
  Then $2x + 1$ = "the sum of twice a number and 1"
  Step 3: *Reread and write an equation*
  The sum of   is −3.

  $\underbrace{\text{twice a number and } 1}$

$$\downarrow \quad\quad\quad \downarrow \quad \downarrow$$
$$2x + 1 \quad\quad = -3$$

  Step 4: *Solve the equation*
  $2x + 1 = -3$
  $2x = -4$
  $x = -2$
  Step 5: *Write your answer*
  The number is −2.
  Step 6: *Reread and check*
  Twice −2 is −4. The sum of −4 and 1 is −3.

15. Step 1: *Read and list*
   Known items: the numbers 6 and 9,
   five times a number.
   Unknown items: the number in question
   Step 2: *Assign a variable and translate*
   Let x = the number asked for
   Then $5x$ = five times the number
   Step 3: *Reread and write and equation*

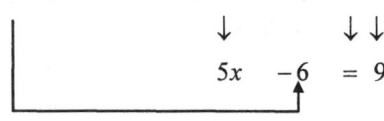

   6 subtracted from five times a number is 9.

   $$5x \quad -6 \quad = 9$$

   Step 4: *Solve the equation*
   $$5x - 6 = 9$$
   $$5x = 15$$
   $$x = 3$$
   Step 5: *Write your answer*
   The number is 3.
   Step 6: *Reread and check*
   Five times 3 is 15. 6 subtracted from 15
   is 9.

17. Step 1: *Read and list*
   Known items: the numbers 1 and 18,
   three times the sum
   Unknown items: the number in question
   Step 2: *Assign a variable and translate*
   Let x = the number asked for
   Then $3(x+1)$ = "three times the sum of
   a number and 1".
   Step 3: *Reread and write an equation*
   Three times   the sum of   is 18.

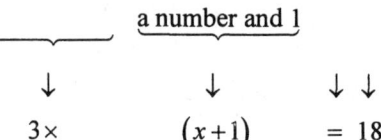

   a number and 1

   $$3\times \qquad (x+1) \qquad = 18$$

   Step 4: *Solve the equation*
   $$3(x+1) = 18$$
   $$3x + 3 = 18$$
   $$3x = 15$$
   $$x = 5$$
   Step 5: *Write your answer*
   The number is 5.
   Step 6: *Reread and check*
   The sum of 5 and 1 is 6. Three times
   that sum is 18.

19. Step 1: *Read and list*

Known items: the numbers 4 and −10, three times a number, five times the sum

Unknown items: the number in question

Step 2: *Assign a variable and translate*

Let x = the number asked for

Then $3x$ = "three times the number".

Step 3: *Reread and write an equation*

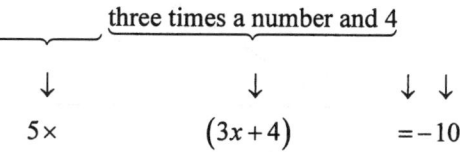

Five times   the sum of   is −10.

three times a number and 4

$\downarrow$ $\downarrow$ $\downarrow$ $\downarrow$

$5\times$ $(3x+4)$ $=-10$

Step 4: *Solve the equation*

$$5(3x+4)=-10$$
$$15x+20=-10$$
$$15x=-30$$
$$x=-2$$

Step 5: *Write your answer*

The number is −2.

Step 6: *Reread and check*

Three times −2 is −6. The sum of −6 and 4 is −2. Five times −2 is −10.

Geometry Problems

21. Step 1: *Read and list*

Known items: Perimeter is 30 m, the length is twice the width

Unknown items: the length and width

Step 2: *Assign a variable and translate*

Let x = width (in meters)

Then $2x$ = length (in meters)

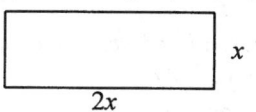

Step 3: *Reread and write an equation*

Perimeter = sum of all sides

$$30 = x+x+2x+2x$$

Step 4: *Solve the equation*

$$30 = x+x+2x+2x$$
$$30 = 6x$$
$$5 = x$$

Step 5: *Write your answer*

The width is $x = 5$ meters.

The length is $2x = 2(5) = 10$ meters.

Step 6: *Reread and check*

The length of 10 m is twice the width of 5 m. The perimeter is 5 + 5 + 10 + 10 = 30 m.

23. Step 1: *Read and list*
Known items: Perimeter is 32 cm
Unknown items: the length of the square
Step 2: *Assign a variable and translate*
Let x = length (in cm) of one side of the square

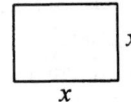

Step 3: *Reread and write an equation*
Perimeter = $4x$

$$32 = 4x$$

Step 4: *Solve the equation*

$$32 = 4x$$

$$8 = x$$

Step 5: *Write your answer*
The length of one side is 8 cm.
Step 6: *Reread and check*
The perimeter is 4(8) cm = 32 cm.

25. Step 1: *Read and list*
Known items: Two equal angles, their sum equals the 3rd angle
Unknown items: the measure of all three angles
Step 2: *Assign a variable and translate*
Let x = measure of 1st angle
Then x = measure of the 2nd angle
and $x + x = 2x$ = measure of the 3rd angle

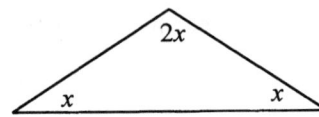

Step 3: *Reread and write an equation*
Sum of all angles = 180°

$$x + x + 2x = 180°$$

Step 4: *Solve the equation*

$$x + x + 2x = 180°$$

$$4x = 180°$$

$$x = 45°$$

Step 5: *Write your answer*
The 1st and 2nd angles equal $x = 45°$.
The 3rd angle is $2x = 2(45°) = 90°$.

Step 6: *Reread and check*
The 3rd angle of 90° is the sum of the two 45° angles.
The sum of the three angles is
$45° + 45° + 90° = 180°$.

27. Step 1: *Read and list*

   Known items: Smallest angle is 1/3 the largest, third angle is twice the smallest

   Unknown items: the measure of all three angles

Step 2: *Assign a variable and translate*

   Let x = measure of largest angle

   Then $\frac{1}{3}x$ = measure of smallest angle

   and $2\left(\frac{1}{3}x\right)$ = measure of the 3rd angle

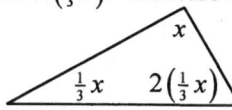

Step 3: *Reread and write an equation*

   Sum of all angles = 180°

   $$x + \frac{1}{3}x + 2\left(\frac{1}{3}x\right) = 180°$$

Step 4: *Solve the equation*

   $$x + \frac{1}{3}x + 2\left(\frac{1}{3}x\right) = 180°$$

   $$x + \frac{1}{3}x + \frac{2}{3}x = 180° \qquad \text{LCD} = 3$$

   $$3\left(x + \frac{1}{3}x + \frac{2}{3}x\right) = 3(180°)$$

   $$3x + x + 2x = 540$$

   $$5x = 540$$

   $$x = 90$$

Step 5: *Write your answer*

   The largest angle is $x = 90°$.

   The smallest angle is

   $$\frac{1}{3}x = \frac{1}{3}(90°) = 30°.$$

   The third angle is

   $$2\left(\frac{1}{3}x\right) = 2\left(\frac{1}{3} \cdot 90°\right) = 60°.$$

Step 6: *Reread and check*

   The smallest angle is 30°, which is 1/3 if 90°. The third angle is 60°, which is twice the smallest angle.

   The sum of the three angles is $30° + 60° + 90° = 180°$.

Age Problems

29. Step 1: *Read and list*

   Known items: Pat is 20 years older than Patrick. In 2 years, the sum of the their ages will be 90.

   Unknown items: Their ages now

Step 2: *Assign a variable and translate*

   Let x = Patrick's age

   Then $x + 20 =$ Pat's age.

Step 3: *Reread and write an equation*

	Now	In 2 years
Patrick	x	$x + 2$
Pat	$x+20$	$(x+20)+2$ $= x + 22$

In 2 years:

   Patrick's age + Pat's age = 90

   $$(x+2) \quad + (x+22) = 90$$

Step 4: *Solve the equation*

   $$(x+2)+(x+22) = 90$$

   $$2x + 24 = 90$$

   $$2x = 66$$

   $$x = 33$$

Step 5: *Write your answer*

   Patrick's age is $x = 33$ years.

   Pat's age is $x + 20 = 33 + 20 = 53$ years.

Step 6: *Reread and check*

   Pat is 53, twenty years older than Patrick, who is 33. In two years, Pat will be 55 and Patrick will be 35, and the sum of those ages is 90.

31. Step 1: *Read and list*
   Known items: Dale is 4 years older than Sue. Five years ago the sum of their ages was 64.
   Unknown items: Their ages now
Step 2: *Assign a variable and translate*
   Let x = Sue's age
   Then $x + 4$ = Dale's age
Step 3: *Reread and write an equation*

	Now	5 years ago
Sue	x	$x - 5$
Dale	$x+4$	$(x+4) - 5 = x - 1$

5 years ago:
Sue's age + Dale's age = 64

$$(x - 5) \ + \ (x - 1) \ = 64$$

Step 4: *Solve the equation*

$$(x - 5) + (x - 1) = 64$$
$$2x - 6 = 64$$
$$2x = 70$$
$$x = 35$$

Step 5: *Write your answer*
   Sue's age is $x = 35$ years.
   Dales's age is $x + 4 = 35 + 4 = 39$ years.
Step 6: *Reread and check*
   Dale is 39 which is 4 years older than Sue, who is 35. Five years ago Dale was 34, Sue was 30, and the sum of those ages is 64.

Renting a Car

33. Step 1: *Read and list*
   Known items: Charges are $10 per day and 16¢ a mile. Total charge is $23.92
   Unknown items: Miles driven
Step 2: *Assign a variable and translate*
   Let x = miles driven
   Then $0.16x$ = cost for miles driven
Step 3: *Reread and write an equation*

$10 per	16 cents	Total
day +	per mile =	cost
10 +	0.16x =	23.92

Step 4: *Solve the equation*

$$10 + 0.16x = 23.92$$
$$10 + 0.16x + (-10) = 23.92 + (-10)$$
$$0.16x = 13.92$$
$$x = 87$$

Step 5: *Write your answer*
   The car was driven 87 miles.
Step 6: *Reread and check*
   The charge for 1 day is $10. The charge for 87 miles is:
   87 miles × $0.16 per mile = $13.92
   The total charge is $10 + $13.92 = $23.92, which checks.

35. Step 1: *Read and list*

Known items: Charges are \$9 per day for 2 days and 15¢ a mile. Total charge is \$40.05

Unknown items: Miles driven

Step 2: *Assign a variable and translate*

Let x = miles driven

Then $0.15x$ = cost for miles driven

Step 3: *Reread and write an equation*

9 per day   15 cents per mile   Total

for 2 days +   for x miles   = cost

$9(2)$ + $0.15x$ = 40.05

Step 4: *Solve the equation*

$$9(2)+0.15x=40.05$$
$$18+0.15x=40.05$$
$$18+0.15x+(-18)=40.05+(-18)$$
$$0.15x=22.05$$
$$x=147$$

Step 5: *Write your answer*

The car was driven 147 miles.

Step 6: *Reread and check*

The charge for 2 days is \$9 × 2 = \$18.
The charge for 147 miles is:
147 miles× \$0.15 per mile = \$22.05
The total charge is \$18 + \$22.05 = \$40.05, which checks.

Coin Problems

37. Step 1: *Read and list*

Known items: She had 10 more dimes than nickels, and a total of \$2.20

Unknown items: Number of dimes and nickels.

Step 2: *Assign a variable and translate*

Let x = number of nickels

Then $x+10$ = number of dimes

	Nickels	Dimes
Number	x	$x+10$
Value	$0.05x$	$0.10(x+7)$

Step 3: *Reread and write an equation*

Value of   Value of   Total

nickels +   dimes   = value

$0.05x$ + $0.10(x+10)$ =   2.20

Step 4: *Solve the equation*

$$0.05x+0.10(x+10)=2.20$$
$$0.05x+0.1x+1=2.2$$
$$0.15x+1+(-1)=2.2+(-1)$$
$$0.15x=1.2$$
$$x=8$$

Step 5: *Write your answer*

She has $x=8$ nickels, and $x+10=8+10=18$ dimes.

Step 6: *Reread and check*

The value of the nickels is $\$0.05(8)=\$0.40$

The value of the dimes is $\$0.10(18)=\$1.80$

The total value is \$0.40 + \$1.80 = \$2.20 which checks.

39. Step 1: *Read and list*

Known items: You have twice as many quarters as dimes. The total value is $9.60

Unknown items: Number of dimes and quarters.

Step 2: *Assign a variable and translate*

Let x = number of dimes

Then $2x$ = number of quarters

	Dimes	Quarters
Number	x	$2x$
Value	$0.10x$	$0.25(2x)$

Step 3: *Reread and write an equation*

Value of   Value of   Total

dimes   +   quarters   =   value

$0.10x$ + $0.25(2x)$ = 9.60

Step 4: *Solve the equation*

$$0.10x + 0.25(2x) = 9.60$$
$$0.1x + 0.5x = 9.6$$
$$0.6x = 9.6$$
$$x = 16$$

Step 5: *Write your answer*

She has $x = 16$ dimes, and

$2x = 2(16) = 32$ quarters

Step 6: *Reread and check*

The value of the dimes is

$0.10(16) = 1.60$

The value of the quarters is

$0.25(32) = 8.00$

The total value is $8.00 + $1.60 = $9.60 which checks.

Miscellaneous Problems

41. Using the 1st column:

$$x + 3 + 4 = 15$$
$$x + 7 = 15$$
$$x = 8$$

Using the 2nd column:

$$1 + 5 + z = 15$$
$$6 + z = 15$$
$$z = 9$$

Using the 3rd column:

$$y + 7 + 2 = 15$$
$$y + 9 = 15$$
$$y = 6$$

43. *Known*: $14/hr for 1st 35 hours, $21/hr over 35 hours, made $574 in 1 week.

Unknown: how many hours she worked

Let x = hours worked

Note: Because $14/hr $\times$ 35 hours = $490, and she made $574, she must have worked more than 35 hours. So x is more than 35. If x is the total hours worked, then $x - 35$ is the number of overtime hours.

	Regular	Overtime
Hours worked	35	$x - 35$
Wage per hour	$14	$21
Total wage	$14(35)$	$21(x - 35)$

Regular wages + overtime wages = $574

$14(35)$ + $21(x - 35)$ = $574

$$14(35) + 21(x - 35) = 574$$
$$490 + 21x - 735 = 574$$
$$21x - 245 = 574$$
$$21x = 819$$
$$x = 39$$

She worked 39 hours.

45. The least amount of money would come from all 36 people being members.

$3 per member $\times$ 36 members = total receipts

$\$3(36) = \$108$

Ike and Nancy receive half the money:

$\frac{1}{2}(\$108) = \$54$

47. *Known*: members pay $3, nonmembers pay $5, 36 people attended, Ike and Nancy received $80.

Unknown: How many of the 36 people were members and how many were nonmembers.

Let x = number of members

Then $36 - x$ = number of nonmembers

	Members	Nonmembers
Number	x	$36 - x$
Cost per lesson	$3	$5
Total money	$\$3(x)$	$\$5(36-x)$

Because Ike and Nancy received $80 which was half the receipts, the total receipts were $160.

Member money + nonmember money = total

$\quad \$3(x) \quad + \quad \$5(36-x) \quad = \$160$

$3x + 5(36-x) = 160$

$3x + 180 - 5x = 160$

$-2x + 180 = 160$

$-2x = -20$

$x = 10$

The number of members was $x = 10$.

The number of nonmembers was

$36 - x = 36 - 10 = 26$.

Yes, it is possible.

Review Problems

49. $4\overline{)3.00}$ with quotient 0.75 $\Rightarrow \frac{3}{4} = 0.75 = 75\%$

$\quad \underline{28}\downarrow$
$\quad\quad 20$
$\quad\quad \underline{20}$
$\quad\quad\quad 0$

51. $1\frac{1}{5} = \frac{6}{5} \Rightarrow 5\overline{)6.0}$ with quotient 1.2 $\Rightarrow 1\frac{1}{5} = 1.2 = 120\%$

$\quad \underline{5}\downarrow$
$\quad 10$
$\quad \underline{10}$
$\quad\; 0$

53. $37\% = \frac{37}{100} = 0.37$

55. $3.4\% = \frac{3.4}{100} = \frac{34}{1,000} = 0.034$

57. What is 15% of 135?

$x = 0.15(135)$

$x = 20.25$

59. 12 is 16% of what number?

$12 = 0.16x$

$\frac{12}{0.16} = \frac{0.16x}{0.16}$

$75 = x$

Section 8.6

1. $A = lw$ *let l = 32, w=22*

$\quad\quad = 32 \cdot 22$

$\quad\quad = 704$

The area is 704 square feet.

3. $A = lw$ *let l = $\frac{3}{2}$, w = $\frac{3}{4}$*

$\quad\quad = \dfrac{3}{2} \cdot \dfrac{3}{4}$

$\quad\quad = \dfrac{9}{8}$

The area is $\dfrac{9}{8}$ square inches.

5. $G = H \cdot R$ *let H = 40, R = 6*

$\quad\quad = 40 \cdot 6$

$\quad\quad = 240$

The gross pay is $240.

7. $G = H \cdot R$ *let H = 30, R = $9\frac{1}{2}$*

$\quad\quad = 30 \cdot 9\dfrac{1}{2}$

$\quad\quad = \dfrac{30}{1} \cdot \dfrac{19}{2}$

$\quad\quad = 15 \cdot 19$

$\quad\quad = 285$

The gross pay is $285.

9. $F = 3 \cdot Y$ *let Y = 4*

$\quad\quad = 3 \cdot 4$

$\quad\quad = 12$

4 yards is equivalent to 12 feet.

11. $F = 3 \cdot Y$ *let Y = $2\frac{2}{3}$*

$\quad\quad = 3 \cdot 2\dfrac{2}{3}$

$\quad\quad = 3 \cdot \dfrac{8}{3}$

$\quad\quad = 8$

$2\dfrac{2}{3}$ yards is equivalent to 8 feet.

13. $I = P \cdot R \cdot T$ *let $P = 1000, R = \frac{7}{100}, T = 2$*

$\quad\quad = 1000 \cdot \dfrac{7}{100} \cdot 2$

$\quad\quad = \dfrac{14,000}{100}$

$\quad\quad = 140$

The interest is $140.

15. $P = 2w + 2l$ *let w = 10, l = 19*

$\quad\quad = 2(10) + 2(19)$

$\quad\quad = 20 + 38$

$\quad\quad = 58$

The perimeter is 58 inches.

17. $P = 2w + 2l$ *let w = $\frac{3}{4}$, l = $\frac{7}{8}$*

$\quad\quad = 2\left(\dfrac{3}{4}\right) + 2\left(\dfrac{7}{8}\right)$

$\quad\quad = \dfrac{3}{2} + \dfrac{7}{4}$

$\quad\quad = \dfrac{6}{4} + \dfrac{7}{4}$

$\quad\quad = \dfrac{13}{4}$

The perimeter is $\dfrac{13}{4}$ feet.

19. $C = \dfrac{5}{9}(F - 32)$ *let F = 212*

$\quad\quad = \dfrac{5}{9}(212 - 32)$

$\quad\quad = \dfrac{5}{9}(180)$

$\quad\quad = 5 \cdot 20$

$\quad\quad = 100$

212°F is equivalent to 100°C (same as in the table).

21. $C = \dfrac{5}{9}(F - 32)$ *let* $F = 68$

$$= \frac{5}{9}(68 - 32)$$

$$= \frac{5}{9}(36)$$

$$= 5 \cdot 4$$

$$= 20$$

68°F is equivalent to 20°C (same as in the table).

23. Since you want to find C, use the formula for C:

$$C = \frac{5}{9}(F - 32) \quad \textit{let } F = 32$$

$$= \frac{5}{9}(32 - 32)$$

$$= \frac{5}{9}(0)$$

$$= 0$$

32°F is equivalent to 0°C.

25. Since you want to find F, use the formula for F:

$$F = \frac{9}{5}C + 32 \quad \textit{let } C = -15$$

$$= \frac{9}{5}(-15) + 32$$

$$= -27 + 32$$

$$= 5$$

−15°C is equivalent to 5°F.

27. The table is as follows:

Age A (yrs)	Max heart rate $M = 220 - A$ (beats per minute)
18	$M = 220 - 18 = 202$
19	$M = 220 - 19 = 201$
20	$M = 220 - 20 = 200$
21	$M = 220 - 21 = 199$
22	$M = 220 - 22 = 198$
23	$M = 220 - 23 = 197$

29. For a 20 year old, the maximum heart rate is $M = 200$ (from problem 27). Therefore, the training heart rate is

$$T = R + \frac{3}{5}(200 - R) = R + 120 - \frac{3}{5}R = \frac{2}{5}R + 120$$

Resting Heart rate, R	Training heart rate $T = \dfrac{2}{5}R + 120$
60	$T = \dfrac{2}{5}(60) + 120 = 24 + 120 = 144$
62	$T = \dfrac{2}{5}(62) + 120 = \dfrac{124}{5} + 120$ $= 144\dfrac{4}{5} \approx 145$
64	$T = \dfrac{2}{5}(64) + 120 = \dfrac{128}{5} + 120$ $= 145\dfrac{3}{5} \approx 146$
68	$T = \dfrac{2}{5}(68) + 120 = \dfrac{136}{5} + 120$ $= 147\dfrac{1}{5} \approx 147$
70	$T = \dfrac{2}{5}(70) + 120 = 28 + 120 = 148$
72	$T = \dfrac{2}{5}(72) + 120 = \dfrac{144}{5} + 120$ $= 148\dfrac{4}{5} \approx 149$

31. a. $6:30 - 2:30 = 4$ hrs

   b. $d = rt = \dfrac{55 \text{ mi}}{\text{hr}} \cdot 4 \text{ hrs} = 220 \text{ mi}$

33. a. $6:30 - 2:30 = 4$ hrs

   b. $r = \dfrac{d}{t} = \dfrac{260 \text{ mi}}{4 \text{ hrs}} = \dfrac{65 \text{ mi}}{\text{hr}}$

35. $V = lwh$ let $l = 6, w = 12, h = 5$

$\quad = 6 \cdot 12 \cdot 5$

$\quad = 360$

The volume is 360 in^3.

37. $V = lwh$ let $l = 6, w = \frac{1}{2}, h = \frac{1}{3}$

$$= 6\left(\frac{1}{2}\right)\left(\frac{1}{3}\right)$$

$$= \frac{6}{6}$$

$$= 1$$

The volume is 1 yd^3.

39. $y = 3x - 2$ let $x = 3$

$\quad = 3(3) - 2$

$\quad = 9 - 2$

$\quad = 7$

41. $y = 3x - 2$ let $x = -\frac{1}{3}$

$$= 3\left(-\frac{1}{3}\right) - 2$$

$\quad = -1 - 2$

$\quad = -3$

43. $y = 3x - 2$ let $x = 0$

$\quad = 3(0) - 2$

$\quad = 0 - 2$

$\quad = -2$

45. $x + y = 5$ let $y = 2$

$\quad x + 2 = 5$

$\quad x = 3$

47. $x + y = 5$ let $y = 0$

$\quad x + 0 = 5$

$\quad x = 5$

49. $x + y = 5$ let $y = -3$

$\quad x + (-3) = 5$

$\quad x = 8$

51. $x + y = 3$ let $x = 2$

$\quad 2 + y = 3$

$\quad y = 1$

53. $x + y = 3$ let $x = 0$

$\quad 0 + y = 3$

$\quad y = 3$

55. $x + y = 3$ let $x = \frac{1}{2}$

$\quad \dfrac{1}{2} + y = 3$

$$y = 3 + \left(-\frac{1}{2}\right)$$

$$y = \frac{5}{2}$$

57. $4x + 3y = 12$ let $x = 3$

$\quad 4(3) + 3y = 12$

$\quad 12 + 3y = 12$

$\quad 3y = 0$

$\quad y = 0$

59. $4x + 3y = 12$ let $x = -\frac{1}{4}$

$$4\left(-\frac{1}{4}\right) + 3y = 12$$

$\quad -1 + 3y = 12$

$\quad 3y = 13$

$$y = \frac{13}{3}$$

61.

$$4x + 3y = 12 \qquad let \ x = 0$$
$$4(0) + 3y = 12$$
$$0 + 3y = 12$$
$$3y = 12$$
$$y = 4$$

63.

$$4x + 3y = 12 \qquad let \ y = 4$$
$$4x + 3(4) = 12$$
$$4x + 12 = 12$$
$$4x = 0$$
$$x = 0$$

65.

$$4x + 3y = 12 \qquad let \ y = -\tfrac{1}{3}$$
$$4x + 3\left(-\frac{1}{3}\right) = 12$$
$$4x + (-1) = 12$$
$$4x = 13$$
$$x = \frac{13}{4}$$

67.

$$4x + 3y = 12 \qquad let \ y = 0$$
$$4x + 3(0) = 12$$
$$4x + 0 = 12$$
$$4x = 12$$
$$x = 3$$

69. The complement of $45°$ is $90° - 45° = 45°$.
The supplement of $45°$ is $180° - 45° = 135°$.

71. The complement of $31°$ is $90° - 31° = 59°$.
The supplement of $31°$ is $180° - 31° = 149°$.

73. a. $S = \dfrac{h \cdot w \cdot fps \cdot t}{35,000}$

$$= \frac{480 \cdot 216 \cdot 30 \cdot 150}{35,000}$$
$$= \frac{466,560,000}{35,000}$$
$$= \frac{466,560}{35} = 13,330\frac{10}{35}$$

The trailer is a little over 13,330 Kb.

b. $S = \dfrac{h \cdot w \cdot fps \cdot t}{35,000}$

$$= \frac{320 \cdot 144 \cdot 15 \cdot 150}{35,000}$$
$$= \frac{103,680,000}{35,000}$$
$$= \frac{103,680}{35} = 2,962\frac{10}{35}$$

The trailer is a little over 2,962 Kb.

75. $A = lw = 6 \cdot 5 = 30$

$P = 30 \qquad N = 4$

$$W = \frac{APN}{2,000}$$
$$= \frac{30 \cdot 30 \cdot 4}{2,000}$$
$$= \frac{3,600}{2,000}$$
$$= \frac{36}{20}$$
$$= \frac{9}{5} \text{ or } 1\frac{4}{5}$$

The weight is $1\frac{4}{5}$ tons.

Note: $\dfrac{9}{5}$ tons $\cdot \dfrac{2000 \text{ lb}}{\text{ton}} = 3,600$ lbs.

Review Problems

77. $\dfrac{\dfrac{3}{5}}{\dfrac{4}{5}} = \dfrac{3}{5} \div \dfrac{4}{5} = \dfrac{3}{5} \cdot \dfrac{5}{4} = \dfrac{3}{4}$

79. $\dfrac{1+\dfrac{1}{2}}{1-\dfrac{1}{2}} = \dfrac{2\left(1+\dfrac{1}{2}\right)}{2\left(1-\dfrac{1}{2}\right)}$

$$= \dfrac{2(1)+2\left(\dfrac{1}{2}\right)}{2(1)-2\left(\dfrac{1}{2}\right)}$$

$$= \dfrac{2+1}{2-1}$$

$$= \dfrac{3}{1}$$

$$= 3$$

81. $\dfrac{\dfrac{1}{2}+\dfrac{1}{4}}{\dfrac{1}{4}+\dfrac{1}{8}} = \dfrac{8\left(\dfrac{1}{2}+\dfrac{1}{4}\right)}{8\left(\dfrac{1}{4}+\dfrac{1}{8}\right)}$

$$= \dfrac{4+2}{2+1}$$

$$= \dfrac{6}{3}$$

$$= 2$$

83. $\dfrac{\dfrac{3}{5}+\dfrac{3}{7}}{\dfrac{3}{5}-\dfrac{3}{7}} = \dfrac{35\left(\dfrac{3}{5}+\dfrac{3}{7}\right)}{35\left(\dfrac{3}{5}-\dfrac{3}{7}\right)}$

$$= \dfrac{21+15}{21-15}$$

$$= \dfrac{36}{6}$$

$$= 6$$

Chapter 8 Review

1. $10x + 7x = (10 + 7)x = 17x$

3. $2a + 9a + 3 - 6 = (2 + 9)a + 3 - 6$
$$= 11a - 3$$

5. $6x - x + 4 = (6 - 1)x + 4$
$$= 5x + 4$$

7. $2a - 6 + 8a + 2 = 2a + 8a - 6 + 2$
$$= (2 + 8)a - 6 + 2$$
$$= 10a - 4$$

9. When $x = 4$:
$$10x + 2 = 10 \cdot 4 + 2$$
$$= 40 + 2$$
$$= 42$$

11. When $x = 4$:
$$-2x + 9 = -2 \cdot 4 + 9$$
$$= -8 + 9$$
$$= 1$$

13. When $x = -3$:
$$5x - 2 = -17$$
$$5(-3) - 2 = -17$$
$$-15 - 2 = -17$$
$$-17 = -17 \text{ true}$$
$x = -3$ is a solution.

15. $x - 5 = 4$
$$x - 5 + 5 = 4 + 5$$
$$x = 9$$

17. $2x + 1 = 7$
$$2x + 1 + (-1) = 7 + (-1)$$
$$2x = 6$$
$$\frac{2x}{2} = \frac{6}{2}$$
$$x = 3$$

19. $2x + 4 = 3x - 5$
$$2x + 4 + (-2x) = 3x - 5 + (-2x)$$
$$4 = x - 5$$
$$4 + 5 = x - 5 + 5$$
$$9 = x$$

21. $3(x - 2) = 9$
$$3x - 6 = 9$$
$$3x = 15$$
$$x = 5$$

23. $3(2x + 1) - 4 = -7$
$$6x + 3 - 4 = -7$$
$$6x - 1 = -7$$
$$6x = -6$$
$$x = -1$$

25. $5x + \dfrac{3}{8} = -\dfrac{1}{4}$ $\qquad$ LCD $= 8$
$$8\left(5x + \frac{3}{8}\right) = 8\left(-\frac{1}{4}\right)$$
$$8(5x) + 8\left(\frac{3}{8}\right) = 8\left(-\frac{1}{4}\right)$$
$$40x + 3 = -2$$
$$40x = -5$$
$$x = -\frac{5}{40}$$
$$x = -\frac{1}{8}$$

27. Let x = number asked for
$\underbrace{\text{The sum of a number and} -3}$ is -5
$$\downarrow \qquad\qquad \downarrow\ \ \downarrow$$
$$x + (-3) \qquad\qquad = -5$$
$$x + (-3) = -5$$
$$x + (-3) + 3 = -5 + 3$$
$$x = -2$$
The number is -2.

29. Let x = number asked for

$\underbrace{3 \text{ times}}$ $\underbrace{\text{the sum of a number and 2}}$ is -6.

$$\downarrow \qquad\qquad \downarrow \qquad\qquad \downarrow\ \downarrow$$

$$3\times \qquad\qquad (x+2) \qquad\qquad =-6$$

$$3(x+2)=-6$$
$$3x+6=-6$$
$$3x=-12$$
$$x=-4$$

The number is –4.

31. Let x = width (in m)
Then $2x$ = length (in m)

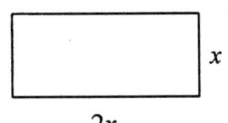

$2x$

P = sum of all sides
$$42 = x + x + 2x + 2x$$
$$42 = 6x$$
$$7 = x$$
The width is $x = 7\,\text{m}$.
The length is $2x = 2\cdot 7 = 14$ m.

33. Let $x = -2$:
$$3x+2y=6$$
$$3(-2)+2y=6$$
$$-6+2y=6$$
$$2y=12$$
$$y-6$$

35. Let $x = 0$:
$$3x+2y=6$$
$$3(0)+2y=6$$
$$0+2y=6$$
$$2y=6$$
$$y=3$$

37. Let $y = 3$:
$$3x+2y=6$$
$$3x+2\cdot 3=6$$
$$3x+6=6$$
$$3x=0$$
$$x=0$$

39. Let $y = 0$:
$$3x+2y=6$$
$$3x+2\cdot 0=6$$
$$3x+0=6$$
$$3x=6$$
$$x=2$$

Chapters 1-8 Cumulative Review

1.
$$5,309$$
$$+\ \ 687$$
$$5,996$$

3.
$$11.090$$
$$-\ \ 6.531$$
$$4.559$$

5.
$$2,305$$
$$\times 407$$
$$16,135$$
$$922,000$$
$$938,135$$

7.
$$\begin{array}{r} 42 \\ 314\overline{)13,188} \\ \underline{1256\downarrow} \\ 628 \\ \underline{628} \\ 0 \end{array}$$

9. *Look at the number to the right of the ten thousands place:*
43**5**,906: $5 \geq 5$ so add 1 to the ten thousands place and replace all digits to the right with 0's: 440,000

11. $\begin{array}{r} 6 \\ 12\overline{)76} \\ \underline{72} \\ 4 \end{array}$ $\Rightarrow \dfrac{76}{12} = 6\dfrac{4}{12} = 6\dfrac{1}{3}$

13. Move the decimal point 2 places to the right and attach a % sign:
$0.8 = 80.0\% = 80\%$

15. $124\% = \dfrac{124}{100} = \dfrac{31}{25} = 1\dfrac{6}{25}$

17. $\left(\dfrac{1}{3}\right)^3 + \left(\dfrac{1}{9}\right)^2 = \dfrac{1}{3}\cdot\dfrac{1}{3}\cdot\dfrac{1}{3} + \dfrac{1}{9}\cdot\dfrac{1}{9}$

$= \dfrac{1}{27} + \dfrac{1}{81}$

$= \dfrac{1\cdot 3}{27\cdot 3} + \dfrac{1}{81}$

$= \dfrac{3}{81} + \dfrac{1}{81}$

$= \dfrac{4}{81}$

19. Because $|-7| = 7$:
$$-|-7| = -7$$

21. $19 - 5(7 - 4) = 19 - 5(3)$
$$= 19 - 15$$
$$= 4$$

23. $\dfrac{3}{8}y = 21$

$\dfrac{8}{3}\left(\dfrac{3}{8}y\right) = \dfrac{8}{3}(21)$

$y = \dfrac{8}{3}(21)$

$y = 56$

25. $\dfrac{3.6}{4} = \dfrac{4.5}{x}$

$3.6(x) = 4(4.5)$

$3.6x = 18$

$x = \dfrac{18}{3.6}$

$x = 5$

27. $5 - (-3) = 5 + 3 = 8$

29. $S = 2lw + 2lh + 2wh$

$= 2(7\text{in})(3\text{in}) + 2(7\text{in})(2\text{in}) + 2(3\text{in})(2\text{in})$

$= 42 \text{ in}^2 + 28 \text{ in}^2 + 12 \text{ in}^2$

$= 82 \text{ in}^2$

31. $\dfrac{432 \text{ miles}}{27 \text{ gallons}} = \dfrac{432}{27} \text{ miles/gal}$

$= 16 \text{ miles per gallon}$

33.

$a = 5 \text{ m}$

$b = 12 \text{ m}$

$c = \sqrt{a^2 + b^2}$

$= \sqrt{5^2 + 12^2}$

$= \sqrt{25 + 144}$

$= \sqrt{169}$

$= 13 \text{ m}$

35. $I = P \times R \times T$

$= \$1,400 \times 0.06 \times \dfrac{90}{360}$

$= \$1,400 \times 0.06 \times \dfrac{1}{4}$

$= \$21$

37. Total checks: $\$376$

$+\ 138$

$\$514$

Opening balance		Total Checks		New Balance
$\$469$	$-$	$\$514$	$=$	$-\$45$

39. Convert gallons to glasses:

15 gal

$= 15 \text{ gal} \times \dfrac{4 \text{ qt}}{1 \text{ gal}} \times \dfrac{2 \text{ pt}}{1 \text{ qt}} \times \dfrac{16 \text{ fl oz}}{1 \text{ pt}} \times \dfrac{1 \text{ glass}}{8 \text{ fl oz}}$

$= \dfrac{15 \times 4 \times 2 \times 16}{8} \text{ glasses}$

$= 240 \text{ glasses}$

Chapter 8 Test

1. $9x - 3x + 7 - 12 = (9 - 3)x + 7 - 12$
$$= 6x - 5$$

3. When $x = 3$:
$$3x - 12 = 3 \cdot 3 - 12$$
$$= 9 - 12$$
$$= -3$$

5. When $x = -1$:
$$4x - 3 = -7$$
$$4(-1) - 3 = -7$$
$$-4 - 3 = -7$$
$$-7 = -7 \text{ true}$$
$x = -1$ is a solution.

7. $a - 2.9 = -7.8$
$$a - 2.9 + 2.9 = -7.8 + 2.9$$
$$a = -4.9$$

9. $\dfrac{7}{x} - \dfrac{1}{6} = 1$
$$\frac{7}{x} - \frac{1}{6} + \frac{1}{6} = 1 + \frac{1}{6}$$
$$\frac{7}{x} = \frac{7}{6}$$
$$7x = 42$$
$$x = 6$$

11. $2(x - 5) = -8$
$$2x - 10 = -8$$
$$2x - 10 + 10 = -8 + 10$$
$$2x = 2$$
$$x = 1$$

13. $6(3x - 2) - 8 = 4x - 6$
$$18x - 12 - 8 = 4x - 6$$
$$18x - 20 = 4x - 6$$
$$18x - 20 + (-4x) = 4x - 6 + (-4x)$$
$$14x - 20 = -6$$
$$14x - 20 + 20 = -6 + 20$$
$$14x = 14$$
$$x = 1$$

15. $r = \dfrac{d}{t} = \dfrac{3{,}100 \text{ mi}}{140 \text{ hr}} \approx 22.14 \text{ or } 22 \text{ mi/hr}$

17. Let x = Susan's age.
Then $x - 5$ is Karen's age.

	Age Today	Age 3 years ago
Susan	x	$x - 3$
Karen	$x - 5$	$(x - 5) - 3 = x - 8$
Sum		$(x - 3) + (x - 8)$

$\underbrace{\text{The sum of their ages 3 years ago}}$ was 11
$$\downarrow \qquad\qquad \downarrow \quad \downarrow$$
$$(x - 3) + (x - 8) \qquad = \quad 11$$
$$(x - 3) + (x - 8) = 11$$
$$x - 3 + x - 8 = 11$$
$$2x - 11 = 11$$
$$2x = 22$$
$$x = 11$$
Susan's age is $x = 11$ years.
Karen's age is $x - 5 = 11 - 5 = 6$ years.